AF589388

Fish in Nutrition

The Authors

Nimish Mol Stephen is currently working as Assistant Professor in Tamil Nadu Dr. M.G.R. Fisheries College and Research Institute, Tamil Nadu Dr. J. Jayalalithaa Fisheries University, Ponneri at the Department of Fish Processing Technology. After having Master Degree in Fisheries Science from Fisheries College and Research Institute, Thoothukudi, she was awarded a Japanese Govt. Scholarship (Monbukagakusho scholarship) to do higher research. She was a Research scholar in the Laboratory of Biofunctional Material Chemistry, Hokkaido University, Japan from 2007 to 2010. She has published four plus research papers in reputed International and National journals. She also published two book chapters in International Publishing groups, and made few research presentations in National and International conferences. She has secured several awards in her credit viz. Justice Ms. M. Fathima Beevi Award for the best MFSc student in Industrial Fish Processing Technology, Dr. J. Jayalalithaa endowment (Gold medal) for the best lady student in MFSc and Dr. J. Jayalalithaa endowment (Gold medal) for the best lady student in BFSc. She is a recipient of Saraswati Scholarship by the Fair & Lovely Foundation, Mumbai and recipient of Student Project Scheme (2002-2003) by the Tamil Nadu State Council for Science and Technology (TNSCST), Chennai. She is also awarded a Young Scientist title by the Venus International Foundation, Chennai (2015).

S Balasundari is working as Professor and Head of the Department of Fish Processing Technology, Tamil Nadu Dr. M.G.R. Fisheries College and Research Institute, Tamil Nadu Dr. J. Jayalalithaa Fisheries University, Ponneri. She handled several research projects and training funded by agencies such as DBT, MPEDA, NFDB, ATMA and ZPD. She has published four research papers in reputed International and National journals. She has presented and participated several National and International conferences and received best poster award. She was awarded a Member of Team, Best KVK for the year 2009.

Dr S Felix, Ph.D., Vice Chancellor, Tamil Nadu Dr. J. Jayalalithaa Fisheries University, Nagapattinam is having experience of more than 35 years in the area of fisheries and aquaculture in teaching, research, extension and administration. He has organized more than 20 Nos. of seminar/ conference/symposium at National and International levels. Currently he is also the President (2018-19) of World Aquaculture Society, Asian Pacific Chapter and Chairman, ICAR's BSMA Committee (2018-19). He has more than 60 research papers published in National and International journals. He has authored more than 10 books and 25 manuals. He is specialized in the area of advanced systems in aquaculture and aquariculture such as raceway, biofloc technology, RAS, etc,. He has operated around 20 externally funded research projects.

Fish in Nutrition

by

Nimish Mol Stephen

S Balasundari

S Felix

Tamil Nadu Dr. M.G.R. Fisheries College & Research Institute
Tamil Nadu Dr. J. Jayalalithaa Fisheries University
Ponneri, Tiruvallur District, Tamil Nadu

DAYA PUBLISHING HOUSE®
A Division of
ASTRAL INTERNATIONAL PVT. LTD.
New Delhi – 110 002

ISBN: 9789388173230 (Int. Edition)

Published by : **Daya Publishing House®**
A Division of
Astral International Pvt. Ltd.
– ISO 9001:2015 Certified Company –
4736/23, Ansari Road, Darya Ganj
New Delhi-110 002
Ph. 011-43549197, 23278134
E-mail: info@astralint.com
Website: www.astralint.com

Laser Typesetting : **Classic Computer Services,** Delhi - 110 035

Printed at : **Neelam Graphics, Delhi - 110007**

डॉ. जे. के. जेना
उप महानिदेशक (मत्स्य विज्ञान)
Dr. J. K. Jena
Deputy Director General (Fisheries Science)

भारतीय कृषि अनुसंधान परिषद
कृषि अनुसंधान भवन-II, पूसा, नई दिल्ली 110 012
INDIAN COUNCIL OF AGRICULTURAL RESEARCH
KRISHI ANUSANDHAN BHAVAN-II, PUSA, NEW DELHI - 110 012
Ph. : 91-11-25846738 (O), Fax : 91-11-25841955
E-mail: ddgfs.icar@gov.in

Foreword

The importance of fish as one of the healthiest diets available in the earth is well accepted today. Fish is not only a source of quality animal proteins, but also unique source polyunsaturated ω-3 fatty acids especially the eicosapentaenoic acid and docosahexaenoic acid, and essential nutrients. Further, the fish is known to be the best source of vitamin D, a nutrient that large mass of the population are deficient in. The therapeutic value in terms of reducing risks of heart attacks, development of brain and eye, and benefits against depression and memory loss have also been well documented. The multiple benefits of fatty fish high in omega-3 contents, and small fish eaten whole, evidently illustrate unique nutritional value of fish. It is quite evident that fish is more affordable than other sources of animal proteins in tropical countries, especially in Indian sub-continent. With the significant growth in both horizontal and vertical expansion in aquaculture during last four decades, the fisheries sector has been playing a big role in meeting the ever-increasing demand of fish and thereby reducing the protein hunger and malnutrition to a great extent in the country and also in several parts of the globe. Nutritional information on different fish species, therefore, would not only help in decision making by the clients on consumption of required fish varieties, but also in prioritizing species for aquaculture.

This book has adequately discussed relevant aspects of '***Fish in Nutrition***' including nutritional composition of fish and shellfish, macro and micro nutrients present in fish, factors affecting nutrients composition, etc.

I congratulate the authors, Drs Nimish Mol Stephen, S. Balasundari and S. Felix for their efforts in bringing out this useful publication, which is expected to be of great value for the students, teachers and also the general consumers at large.

(J.K. Jena)

Message

"Fish in Nutrition" is a fundamental course of Fish Processing Technology which is being taught in the curriculum of graduate programme in Fisheries Science. This book gives a comprehensive account of nutrients in fish and their benefits. It includes detailed information about major and minor nutritional components of fish and shellfish including lipids, proteins, vitamins and minerals. This book introduces students to the individual major and minor nutrients of fish and shellfish, benefits of its consumption and the recommended daily intake levels. An effort has been made to provide collective information for the benefit of students of Fisheries Universities and researchers pursuing careers in fisheries, food, nutrition and allied courses.

S. Felix

Vice-Chancellor

Tamil Nadu Dr. J. Jayalalithaa Fisheries University

Preface

Fish and nutrition aids becoming an increased focus on both developing and developed countries. Fish is the major source of animal protein in several developing countries, which provide essential micronutrients to vulnerable populations. Fish also serve as a solution to several existing health problems especially cardiovascular diseases through its vital functional nutrients. Currently, the dietary patterns are shifting in developed and middle-income countries adding increased amount of fish in their food culture.

Fish, the "nature's unique super food" is surpassingly rich in essential nutrients, including long-chain omega-3 fatty acids, iodine, vitamin D, and calcium. The study of fish in human nutrition is accomplishing omnipresence across the world as the role of fish and its benefits is components indisputable. The study of fish in human nutrition needs a solid base in the compositional architecture of fish and the effect of different processing methods on it. The basis of this textbook is to provide collective information from different sources in one folder as Fish in Human Nutrition.

This textbook comprises of eleven chapters which covers right from the nutritional composition of fish and shellfish to the effect of different cooking methods on nutritional composition of fish. It provides a detailed note on the macro and micro nutrients present in fish and also the factors affecting their composition.

Authors

Contents

Chapter 1

Composition of Fish

1. Introduction

Biochemical composition is defined as chemical composition of a particular food. It includes macronutrients and micronutrients. Macronutrients include carbohydrate, protein, fat, fibre and water. Micronutrients includes vitamins, minerals, flavonoids, carotenoids and polyphenols etc. Food composition data (FCD) are detailed sets of information on the nutritionally important components of foods and provide values for energy and nutrients including protein, carbohydrates, fat, vitamins, minerals and fibre.

1.1. Why Biochemical Composition of Fish is Important?

Fish is one of the most valuable sources of high grade protein and lipid. Customer awareness of its composition is essential to utilize it to the fullest. Knowing the nature of raw material will help

- ☆ to choose the processing techniques
- ☆ to know the contribution of fish to the diet and to health
- ☆ to know the taste of fish, nutritional content of fish
- ☆ lean or fatty in order to prepare different dishes for the table
- ☆ to know the composition of whole fish for fish meal manufacture
- ☆ to know the oil content of liver and muscle by fish oil manufacturer
- ☆ Measurement of constituents of fish products is necessary to meet specifications or to comply with regulations.

1.2. What is Proximate Analysis of Foods?

This system of analysis divides the food into six fractions: moisture, ash, crude protein, ether extract/crude lipid, fibre and nitrogen-free extractives. The etymology of the words-proximate composition – is uncertain.

The moisture content is determined as the loss in weight that results from drying a known weight of food to constant weight at 100°C. This method is satisfactory for most foods, but with a few, such as silage, significant losses of volatile material may take place.

The ash content is determined by ignition of a known weight of the food at 550°C until all carbon has been removed. The residue is the ash and is taken to represent the inorganic constituents of the food. The ash may, however, contain material of organic origin such as sulphur and phosphorus from proteins, and some loss of volatile material in the form of sodium, chloride, potassium, phosphorus and sulphur will take place during ignition. The ash content is thus not truly representative of the inorganic material in the food either qualitatively or quantitatively.

The crude protein (CP) content is calculated from the nitrogen content of the food, determined by a modification of a technique originally devised by Kjeldahl over 100 years ago. In this method the food is digested with sulphuric acid, which converts all nitrogen present except that in the form of nitrate and nitrite to ammonia. This ammonia is liberated by adding sodium hydroxide to the digest, distilled off and collected in standard acid, the quantity so collected being determined by titration or by an automated colorimetric method. It is assumed that the nitrogen is derived from protein containing 16 percent nitrogen, and by multiplying the nitrogen figure by 6.25 (*i.e.* 100/16) an approximate protein value is obtained. This is not 'true protein' since the method determines nitrogen from sources other than protein, such as free amino acids, amines and nucleic acids, and the fraction is therefore designated crude protein.

The ether extract (EE) fraction is determined by subjecting the food to a continuous extraction with petroleum ether for a defined period. The residue, after evaporation of the solvent, is the ether extract. As well as lipids it contains organic acids, alcohol and pigments. In the current official method, the extraction with ether is preceded by hydrolysis of the sample with sulphuric acid and the resultant residue is the acid ether extract.

The carbohydrate of the food is contained in two fractions, the crude fibre (CF) and the nitrogen-free extractives (NFE). The former is determined by subjecting the residual food from ether extraction to successive treatments with boiling acid and alkali of defined concentration; the organic residue is the crude fibre.

When the sum of the amounts of moisture, ash, crude protein, ether extract and crude fibre (expressed in g/kg) is subtracted from 1000, the difference is designated as nitrogen-free extractives. The crude fibre fraction contains cellulose, lignin and hemicelluloses. The nitrogen-free extractive includes sugars, fructans, starch, pectins, organic acids and pigments.

2. Proximate Composition of Fish

The percentage composition of the four major constituents of fish viz. water, protein, lipid and ash (minerals) account for about 96-98 percent of total tissue constituents in most cases. Carbohydrates, vitamins, nucleotides, other non-protein nitrogenous compounds etc. are also present in small quantities.

The major feature of proximate composition of fish and shellfish is its great variability in lipid content. Flesh and edible parts contain many minor components such as glycogen, sugar, sugar phosphates, nucleotides, water soluble and fat soluble vitamins, cholesterol, haemoglobin, myoglobin, other pigments, non protein nitrogenous substances (such as free amino acids, urea predominantly in cartilaginous fishes, trimethylamine oxide mainly in marine fish and crustaceans), enzymes, hormones, sterols, phospholipids, occasionally hydrocarbons and many other substances. Though quantitatively minor components, they play vital roles in maintaining the system and thus are essential for growth and development of the organisms.

Table 1.1: Range of Proximate Composition in Edible Portion of Common Fishes from Indian Coastal Waters

Water	65-90 percent
Protein	10-22 percent
Fat	0.5-20 percent
Mineral	0.5-5 percent

2.1. Significance of Chemical Composition of Fish

Edible parts of a fish and shellfish gain importance than whole fish in food chemical composition data. Edible part differ with different regions of the world.

The compositions of flesh and its structure determine the organoleptic qualities (such as odour and flavour, appearance and texture), palatability as a food, freezing behaviour, storability both at ambient and cold temperature spoilage pattern.

Fish and shellfish have a few specific physiologic and biochemical features which make them distinct and separate from land animals:

- ☆ Poikilotherm nature
- ☆ Presence of trimethylamine oxide (TMAO)
- ☆ High content of non-protein nitrogen (NPN)
- ☆ Variable lipid content with high degree of unsaturation and high amount of omega-3 fatty acids

☆ Low carbohydrate content causing comparatively a high post-mortem pH in the flesh (usually 6.0).

Above biochemical specialties make a fish food product more vulnerable to microbial spoilage and deterioration due to the development of rancidity.

2.2. Moisture Content of Fish

Bombay duck fish (*Harpodon nehereus*) whether it is a catch from the Arabian Sea or the Bay of Bengal has the peculiarity of having very high moisture content (about 90 percent) and low protein content (about 10 percent). It is lean fish with < 1 percent fat in its flesh.

2.3. Fat

Major components of lipids from fish are triglycerides, phospholipids, cholesterol and cholesteryl esters. In skeletal muscle of fish, fat soluble viatmins are present in small amounts. Liver oils of cod, halibut and shark are rich in vitamin A and/ or vitamin D. Fish oil commercially extracted from fish or its offal is composed chiefly of triglycerides with small quantities of other components phospholipids and various unsaponifiable matters (not exceeding 3 percent of oil). Liver oils of elasmabranch fish often contain higher percentage of unsaponifiable matter (quite often 8-14 percent), sometime as high as 80 percent.

2.4. Proteins

Proteins are organic nitrogenous substances formed from subunits called amino acids. Protein is the second major component in fish muscle. Proteins of fish muscle are classified into three groups

a. Sarcoplasmic protein

Protein soluble in salt solutions of low ionic strength(<0.15) which accounts for 25-30% of the total protein. These include myogen and globulin. It is high in pelagic fish, while low in demersal fish

b. Myofibrillar protein

Protein soluble in solutions of high ionic strength (>0.5) which accounts for 65% of the muscle protein. These include actin, myosin, actomyosin, tropomyosin and troponin. Gelling property essential for surimi production is based on this protein.

c. Stroma protein

Protein insoluble in neutral salt solutions or in dilute acids or alkalies, which accounts for 3-10% of fish. Collagen is another form of protein similar to stroma protein.

3. Factors Influence Proximate Composition of Fish

Fishes are a very heterogenus group. The vast number of investigation carried out in different parts of the world indicate that the proximate and the chemical composition of a fish depends up on the genus and the species. Again in the same species, the composition varies with age and size, sex and sexual maturity, habitat and fishing season.

3.1. Anatomical differences

Distribution of different constituents in the body is uneven. Different organs such as flesh, skin, scales, head, bones, fins, liver, and other parts of viscera have different composition.

- ☆ Skin portion of many fatty or medium fatty fishes has more fat or lipid content compared to fish muscle
- ☆ In lean fish, the difference in fat contents of skin and flesh is only marginal
- ☆ Fish offal (mainly head and viscera parts) of oil sardine contain more fat than beheaded and eviscerated part.
- ☆ Viscera of freshwater murrels have more lipid content compared to that of flesh
- ☆ Skeletal muscle of different parts of the body shows considerable variations
- ☆ In lean fish, Wallago attu there is a decline in fat content from tail to head and from ventral to dorsal.
- ☆ The flesh of dorsal or posterior region has more moisture and less fat content compared to ventral or anterior region
- ☆ Quality of proteins of skin is inferior to muscle
- ☆ Body meat and claw meat of crab do not show significant difference
- ☆ Body meat of crab contain less pigment than claw meat, this gives more appealing appearance.

Table 1.2: Chemical Composition of White Pomfret (*Stromateus cinereus*)

Sl.No.	*Fish*	*Skin*	*Muscle*	*Whole*
1.	Proportionate weight from beheaded eviscerated, deboned fish	5	95	100
2.	Moisture (percent)	63.0	75.0	74.6
3.	Total extractable proteins (percent)	1.8	15.8	15.1
4.	Sarcoplasmic proteins (percent)	0.6	3.7	3.5
5.	Fibrillar proteins (percent)	1.2	12.2	11.6
6.	Total extractable lipids (percent)	16.2	4.3	4.9
7.	Phospholipids (percent)	2.5	1.0	1.0
8.	Neutral lipids (percent)	11.0	2.4	2.8
9.	Free fatty acids (percent)	0.4	0.1	0.1

3.2. Red and White Muscle

Colour or appearance of muscle of most fish is not uniform. It is often coloured brown or reddish in places especially along the body immediately under the skin (subcutaneous muscle), while the muscle that forms a fish fillet is generally white to off-white in colour. The brown or reddish coloured muscle is called dark or red muscle. Dark muscle may also be present near the spine.

Finfish differs from terrestrial animals by the presence of dark muscle. Ratio of dark red meat to white meat varies from species to species. In dark fleshed species the ratio is 9 to 24 percent, 4 to 8 percent ranges for species in intermediate position and white flesh species are 2 to 3 percent. The content of red meat expressed as percentage of the total muscle ordinarily shows an increase with age. Shoaling fishes especially of the scombroid group contain appreciable amounts of red meat in their muscle. In small fishes like sardine it amounts to about 37 percent of the total muscle whereas in big fishes like tuna it is only to the extent of 10-12 percent.

- ☆ Dark muscle contains more (about 5 times) of hemepigments, particularly myoglobin and haemoglobin compared to white muscle. These pigments give the dark muscle its characteristic colour and appearance.
- ☆ Dark muscles are required for slow and continuous movement such ascontinuous swimming while white muscles for quick and occasional activity and help in rapid energy supply.
- ☆ Demersal gadoid fish which often rest in the sea-bed bottom has much less dark muscle than in the pelagic fish which swim throughout the life.
- ☆ Dark muscles also contain more of mitochondria and less of sacroplasmic reticulum than white muscles.
- ☆ Dark muscles contain larger concentrations of lipids (2 to 5 times), B-vitamins, cytochrome C, glycogen, reducing sugars and nucleic acid than white muscles.
- ☆ White muscles contain more of water, protein, inorganic phosphates, lactic acid, liver lipase, ATPase and glycolytic enzymes.
- ☆ Glycogen is the chief fuel of white muscle contraction.
- ☆ Lipid is the chief fuel of dark muscle

In case of tuna, oil sardine and *Cirrhinus mrigala*, red meat contains more fat than white meat. However, with hilsa (*Hilsha ilisha*) fish, dark muscle appears to have less lipid than whole fish or white muscle from dorsal and ventral parts.

- ☆ Essential amino acids are more in red meat than in white meat of oil sardine (*Sardinella longiceps*).
- ☆ Free amino acid profiles of white and red meat of tuna (*Katsuwonus pelamis*) indicate that white meat contains more of all the amino acids excepting histidine and serine which are more in red meat.

- ☆ Red meat appears to contain more of iron and less of calcium and potassium compared to white meat.
- ☆ Sodium contents of both the meats are more or less the same.
- ☆ Protein efficiency ratio (PER value) shows that red meat of oil sardine fish is marginally more nutritious than white meat and casein.
- ☆ Dark fleshed species are not suitable for surimi production, it may give an unattractive mince.
- ☆ In canned tuna, only white meat separated from dark meat is processed.

Table 1.3: Proximate Composition and Chemical Characteristic of Red and White Meat of Marine Fish Tuna (*Euthynnus affinis*) and Oil Sardine (*Sardinella longiceps*) (All figures are on wet weight basis)

	Tuna		*Oil Sardine*	
	Red Meat	*White Meat*	*Red Meat*	*White Meat*
Moisture (percent)	69.4	70.9	64.6	72.8
Protein (total N×6.25)	18.3	18.9	19.1	20.7
Fat (percent)	4.6	3.1	14.1	4.8
Ash (percent)	1.2	1.7	1.4	1.7
Carbohydrates (percent)	0.8	0.3	0.9	0.5
Reducing sugars (mg/100g)	110	10.6	-	-
Lactic acid (mg/100g)	23.0	44.0	-	-
Inorganic phosphate (mg/100g)	120	124	306	370
Sodium (mg/100g)	53.7	47.5	149	131
Potassium (mg/100g)	238.4	391.2	361	404
Calcium (mg/100g)	134.4	178.8	103	112
Iron (mg/100g)	11.1	4.8	11	4

3.3. Size and Age

- ☆ Increases in size or length of species of fish by age is decided by genetic factor.
- ☆ After reaching maximum attainable size or length, it may age without further increase in size (length).
- ☆ Ordinarily fully grown fish should constitute the major fraction of a commercial catch and variation in chemical composition for such catches of a species is not much.
- ☆ However, in commercial landings of many species, fishes of different sizes (lengths) apparently with different ages are often landed and are sold either for use as a food or as a raw material for feed.
- ☆ With many freshwater species, fish of different sizes of the same species forms a commercial commodity for sale

3.4. Sex

- ☆ Variation in proximate and chemical composition of a species originates from the development of reproductive organs namely gonads of male and female.
- ☆ The developed gonads of a male or a female member of a species have different composition. For example, in a freshwater minor carp, male gonad has less protein and more water than female gonad.
- ☆ The difference in the proximate composition of the flesh of male and female members of a species during spawning season is substantial.

3.5. State of Maturity

- ☆ Protein and lipid contents of muscle, liver and gonads are in peak values at maturing and ripening stages for both male and female of freshwater fish. Eg. Carp (Cirrhinus mrigala) and marine fish (Otolithes argenteus).
- ☆ With many marine species of Atlantic Ocean, protein content shows certain reduction by about 2 percent in female during sexual cycle but this phenomenon is not observed in male.

3.6. Seasonal Variation

Significance of seasonal variation in the proximate composition which has been observed both in marine and in freshwater species. The most important factors are :

(i) Reproductive development

Life-cycle of an adult fish may be divided into two basic periods: (a) fish goes through a period that is concerned with reproduction (gonad development, migration and spawning) and (b) fish lives through a phase of intense feeding or recovery from spawning which lasts until the gonads begin to develop once again. During the development of gonads and spawning, both freshwater and marine fish expends a great deal of energy with fat as its principal source. During the period of recovery after spawning, body-reserves which were lost as a result of spawning are built-up, and fish starts accumulating fat again.

(ii) Feeding conditions

Quantity and quality of food available

- Fat and moisture contents are found to be interrelated. Seasonal variations in the content of other nitrogenous and mineral substances are usually less marked and are, therefore, seldom considered in estimates.
- Oil sardine (*Sardinella longicep*) has the maximum fat content and minimum moisture content during September – January period. The change in the triglyceride content generally follows the pattern of total lipid (fat), but the phospholipid content remains fairly constant at about 1 percent level throughout the year.

Chapter 2

Concept of Biological Value, Protein Efficiency Ratio and Net Protein Utilization of Energy Nutrients

1. Introduction

Energy is provided by oxidation of the major nutrients. The first requisite of an adequate diet is a source of energy, provided by oxidation of the three bulk nutrients: carbohydrate, fat and protein.

The unit of energy used in calculating energy requirements is the kilocalorie (Kcal), also called the nutritional calorie. Gram calorie is the amount of energy required to raise the temperature of 1.0 g of water by 1°Celsius.

The amount of energy released by oxidation of carbohydrates, fats, and proteins has been determined by burning weighed samples in an atmosphere of oxygen in a bomb calorimeter and measuring the total amount of heat produced.

Energy yield by different source

Carbohydrate : 4.2 Kcal/g,

Fats : 9.5 Kcal/g

Proteins : 4.3 Kcal/g.

When oxidized in the body, foods are not completely digested and absorbed; and hence yield an amount of heat is not equal to the heat released when they are oxidized in a calorimeter.

Table 2.1: Energy Needs for Various Activities

Function	*Activity*	*Men Kcal/kg/h*	*Women Kcal/kg/h*
Very light	Seated and standing activities, painting, auto and truck driving, laboratory work, typing	1.5	1.5
Light	Walking (2.5-3, miles/h), carpentry, shopping restaurant trades, washing clothes, glf	2.9	2.6
Moderate	Walking fast (3.5-4, miles/h), jogging weeding, hoeing, cycling, tennis, dancing, volleyball	4.3	4.1
Heavy	Walking with load uphill, sawing wood, pick and shovel labour, swimming, climbing, foot ball.	8.4	8

Calorific value

The amount of energy produced by the complete combustion of a material or fuel. Measured in units of energy per amount of material, example. kJ/kg (1kJ=0.239kcal). The amount of energy available from a food item when digested, mostly from carbohydrates and fats.

2. Calorific value of Food

The energy accumulated in food substances (Proteins, fat and carbohydrates) are expressed in calories (cal) or kilocalories (kcal). This concept is used in the comparative evaluation of food products and in planning diets.

2.1. Carbohydrate

Carbohydrates are the major source of energy. Carbohydrate rich foods are abundant and cheap compared with fats and protein. They naturally form major part of the diet in most of the world. They occur in food as sugars and starches, which are the source of energy. The simple sugars are sweet in taste. Glucose is the simple unit from which more complex carbohydrates are built.

2.2. Fat

Fats provide calories and essential fatty acids.

- ☆ Triacyglycerols constitute about 98 percent of total dietary lipids, the remaining 2 percent consists of phospholipids and cholesterol and its esters.
- ☆ At room temperature, triacylglycerols of animal origin, which contain a relatively large proportion of saturated fatty acids, are usually solid and those of plant origin, which contain a greater fraction of unsaturated fatty acids, are usually liquid.
- ☆ On oxidation in the tissues triacylglycerols of both types provide over twice as much energy per gram as carbohydrates.

- Fats tend to remain in the stomach longer than carbohydrates and are digested more slowly, they also have greater satisfy value than carbohydrates.
- Most animal fats (from meat, milk and eggs) are relatively rich in saturated fatty acids and low in polyunsaturated fatty acids, except chicken and fish.
- Fish fat in general consist primarily of triglycerides, esters of fatty acids,less amount of free fatty acids, sterols, vitamins and hydrocarbons. They resemble vegetable oils and animals fats in having triglyceryl esters of fatty acids but they differ in having a variety of fatty acids of chain length, from C14 to C24.
- Fish fat is rich in polyunsaturated fatty acids, especially omega-3 fatty acids.
- The principal fatty acids of animal and vegetable fats contain 2 to 3 double bonds per molecule whereas fish oils contain fatty acids with 4 to 5 double bonds.
- Low density lipoprotein (LDL) is also called as bad cholesterol. High density lipoprotein (HDL) is also called as good cholesterol, as it play a cholesterol removing carrier role in body.
- Consumption of diet rich in saturated fats tend to decrease the concentration of HDL. There is a positive correlation between the incidence of coronary heart diseases and low levels of HDL and also high levels of LDL and total cholesterol. Saturated animal fat should be replaced in part by a balanced diet rich in polyunsaturated fatty acids.

Cholesterol in the diet also appears to affect the proportions of the lipoproteins in the blood in some individuals. Cholesterol is present in significant amounts in animal products, particularly egg yolk, butter fat, and meat, but is absent in plant foods. Although a diet high in cholesterol will increase the blood cholesterol level, dietary cholesterol at the same time inhibitory to the biosynthesis of cholesterol in the tissues. There is a delicately regulated balance between the amount of ingested cholesterol and the amount of cholesterol synthesized in the body protein. The excess would be deaminated in the liver and converted into glycogen or fat or burned as fuel.

2.3. Protein

Proteins are required for their amino acid content. Proteins also provide essential amino acids which are not synthesized in the body. The diet contains a variety of different animal and plant proteins.

The nutritional value or quality of a given protein depends upon two factors: (1) content of the essential amino acids and (2) digestibility.

Some proteins contain a complete set of essential amino acids in a proper proportions; others may be deficient in one or more essential amino acids. Example: Cereals are low in lysine, while pulses are low in methionine.

Plant proteins, particularly those of wheat and other grains, are not completely hydrolysed during digestion because the protein-rich portions of the grains are surrounded by protective husks of cellulose and other polysaccharides that are not hydrolyzed by intestinal enzymes. Since only free amino acids can be absorbed from the intestine, not all the amino acids of most plant foods is biologically available.

3. Nutritional Quality of Protein

The nutritional quality of proteins can be determined by different methods.

3.1. Biological Value [BV]

A more accurate measure of quality of protein, which describe how the amount of a given protein that must be consumed to keep an adult human or experimental animal with nitrogen balanced. The condition in which the intake of protein nitrogen exactly balances the loss of nitrogen in the urine and feces. BV is an index of protein quality that reflects the percentage of absorbed nitrogen from dietary protein actually retained by the body under standard conditions

$$BV = \frac{\text{N retained}}{\text{N absorbed}} \times 100$$

For example, if rats are fed on a controlled diet using a given protein food source, the nitrogen content of the diet, urine and faeces is measured and BV is calculated as:

$$BV = \frac{\text{Dietary N} - (\text{Urinary N} + \text{Faecal N})}{\text{Dietary N} - \text{Faecal N}} \times 100$$

- ☆ If a given protein provides all the essential amino acids in proper proportions and all are released in free form and absorbed, it will have a biological value of 100.
- ☆ This value is much higher than a protein that is complete in amino acid content but is incompletely digested or one that is completely digested but is low in one or more essential amino acids.
- ☆ According to this test, if a given protein is totally deficient in only one essential amino acid, it will have zero biological value.
- ☆ If a protein has a very low biological value, a very large amount must be consumed in order to provide the minimal requirement of whatever essential amino acids it provides in least amount.
- ☆ Protein would be consumed in excess of the amount needed for synthesis of body protein. The excess would be deaminated in the liver and converted into glycogen or fat or burned as fuel.
- ☆ Animal proteins, for example, those of milk, fish, beef steak, and eggs rank high (>80) in both chemical score and biological value.
- ☆ Proteins of rice or wheat have a low chemical score because they are

deficient in one or more essential amino acids. However, they also have lower biological value, since they are incompletely digested.

☆ Thus, a much larger amount of these plant proteins must be consumed to provide the minimum daily requirement of all the amino acids.

3.2. Chemical Score

Upon complete hydrolysis of protein, its amino acid composition is measured and compared with that of egg or milk protein as a standard. The chemical score of a protein indicates the potential value of the protein. The chemical score is calculated as

$$\text{Chemical score} = \frac{\text{mg of amino acid per gram of test protein}}{\text{mg of amino acid per gram of reference protein}} \times 100$$

3.3. Net Protein Utilization [NPU]

NPU is an index that takes into account of the relative digestibility of proteins. Because, even the best mixture of amino acids will be less available for use in the body if it is packaged in a protein that is only partially digested. Therefore, NPU = BV * digestibility.

$$BV = \frac{\text{N retained}}{\text{Dietary N}} \times 100 = \frac{\text{Dietary N} - (\text{Urinary N} + \text{Faecal N})}{\text{Dietary N}} \times 100$$

Proteins are generally easy to digest. Most proteins are 90 percent or more digestible. Thus in most cases, NPU approximates the BV.

3.4. Protein Efficiency Ratio [PER]

This index is not based on nitrogen balance studies. It is, therefore, less precise than BV\NPU, but, it is technically easier to derive and use. PER is defined as the change in body weight relative to the amount of protein consumed.

Table 2.2: Digestibility Coefficients and Biological Values of different Fishes at 10 percent Protein in take

Name of Fish	*Digestibility Coefficients*	*Biological Value*
Catla catla	86.45	77.96
Cittihus mrigala	92.38	72.16
Labeo rohita	88.6	78.9
Hilsa ilisha	82.6	69.5
Formio niger	97.14	91.95
Mugil speigleri	---	82.21
Coliia dussmieri	95.44	92.04
Harpodon nehereus	90.47	95.27
Scoliodon sorrak owah	75.42	54.13
Rhynchobatus djiddensis	72.1	60.1

$$PER = \frac{\text{Weight gain in grams}}{\text{Dietary protein in grams}} \times 100$$

For example, if a rat is given a standard diet, containing 2g of casein per day as the only source of protein and the weight gain is found to be 5g per day, then the PER of casein would be 5/2=2.5. Whole egg has a PER of 3.8, while gelatin has 0 PER.

4. Deficiency Diseases

Two forms of child undernutrition, often occurring together, are Marasmus and Kwashiorkor.

Marasmus

Marasmus (from Greek, "to waste") is the term applied to chronic deficiency of calories in children; it is caloric starvation. Marasmus occurs in famine areas when infants are given inadequate bottle feedings of thin watery gruels of native cereals or other plant foods, usually deficient in both calories and protein. Marasmus is characterized by arrested growth, extreme muscle wasting, weakness, and anemia. It is usually complicated by multiple deficiencies of vitamins and minerals. Calorie deficiency in early childhood, even if ultimately alleviated with an ample diet, leaves a permanent deficit in body growth.

Kwashiorkor

Chronic protein deficiency in children is called "Kwashiorkor", an African word that means "weaning disease". The growth of protein deficient children is retarded, they become anemic, and the tissues become watery and bloated because of the low serum protein levels, which upset the normal distribution of water between the tissues and the blood. Moreover, the liver, kidneys, and pancreas undergo severe degeneration. The mortality rate of kwashiorkor is very high. There is a permanent physiological deficit.

5. Caloric over Nutrition

Obesity is the result of caloric over nutrition. Obesity, which increases the risk of cardiovascular diseases, hypertension, and diabetes, is quite simply the result of caloric intake in excess of body needs. It usually begins in childhood or adolescence, and the longer it is allowed to persist, the less likely that it can be controlled.

Chapter 3

Amino Acids of Fish and Shell Fishes

1. Introduction

Proteins, the most abundant macromolecules found in biological systems, are present in diverse forms such as structural elements, enzymes, hormones, antibodies, receptors, signaling molecules, etc., having specific biological functions. Protein is necessary for key body functions including provision of essential amino acids, development and maintenance of muscles.

- ☆ Amino acids serve as building blocks of proteins
- ☆ Serve as intermediates in various metabolic pathways
- ☆ They serve as precursors for synthesis of a wide range of biologically important substances including nucleotides, peptide hormones, and neurotransmitters
- ☆ Amino acids play important roles in cell signaling
- ☆ They act as regulators of gene expression and protein phosphorylation
- ☆ They act as regulators of cascade nutrient transport and metabolism in animal cells
- ☆ They act as regulators of innate and cell-mediated immune responses.

2. Classification

Based on the nutritional requirements, amino acids are grouped into different classes:

1. Essential (EAA)
2. Nonessential (NEAA)

3. Conditionally essential Amino Acids
4. Functional amino Acids
5. Semi-essential amino Acids

a. Essential (EAA)

- ☆ Amino acids which cannot be synthesized by the body to meet the biological needs
- ☆ Need to be supplied through diet
- ☆ They are required for proper growth and maintenance of individual.

Example: Arginine, cystine, histidine, leucine, lysine, methionine, threonine, tryptophan, tyrosine, and valine.

b. Nonessential (NEAA)

- ☆ Non essential Amino acids are synthesized in the body to meet the biological needs
- ☆ They need not be supplied through diet
- ☆ They play important roles in regulating gene expression and micro-RNA levels.
- ☆ They play important roles in regulating cell signalling, blood flow, nutrient transport and metabolism in animal cells.
- ☆ They play important roles in the development of brown adipose tissue
- ☆ They play important role in the intestinal microbial growth and metabolism
- ☆ They are also involved in anti-oxidative responses, and innate and cell-mediated immune responses.

Example: Aspartic acid, serine, and alanine are NEAA.

c. Conditionally Essential Amino Acids

Glutamine, glutamic acid, glycine, proline, and taurine are CEAA.

d. Functional Amino Acids

- ☆ FAA are those which participate and regulate key metabolic pathways to improve health, survival, growth, development and reproduction of the organisms.
- ☆ They play a key role in prevention and treatment of metabolic diseases (*e.g.*, obesity, diabetes, and cardiovascular disorders), intrauterine growth restriction, infertility, intestinal and neurological dysfunction, and infectious disease.

Example: Arginine, cystine, leucine, methionine, tryptophan, tyrosine, aspartate, glutamic acid, glycine, proline, and taurine are FAA.

e. Semi-essential Amino Acids

Arginine and histidine can be synthesized by adults and not by growing children. Hence, they are considered as semi-essential amino acids

3. Role of Amino Acids in Human Health

Amino acids are mainly obtained from proteins in diet and the quality of dietary protein is assessed from essential to nonessential amino acid ratio. High quality proteins are readily digestible and contain the dietary essential amino acids (EAA) in quantities that correspond to human requirements. Inadequate uptake of quality proteins and calories in diet leads to protein energy malnutrition (PEM) (or protein-calorie malnutrition, PCM) which is the most lethal form of malnutrition/hunger. Kwashiorkor and marasmus, the extreme conditions of PCM mostly observed in children, are causes by chronic deficiency of protein and energy, respectively. PCM also occurs in adults who are under chronic nutritional deficiency.

a. Arginine

Arginine play an important role in cell division, wound healing, ammonia removal, immune function, and hormone release. It is also the precursor for biological synthesis of nitricoxide which play important roles in neurotransmission, blood clotting, and maintenance of blood pressure. It is supplemented for recovery from a number of diseases like sepsis, preeclampsia, hypertension, erectile dysfunction, anxiety, and so forth.

b. Leucine

Leucine is the only dietary amino acid that can stimulate muscle protein synthesis and has important therapeutic role in stress conditions like burn, trauma, and sepsis. As a dietary supplement, leucine has been found to slow the degradation of muscle tissue by increasing the synthesis of muscle proteins.

c. Methionine

Methionine is used for treating liver disorders, improving wound healing, and treating depression, alcoholism, allergies, asthma, copper poisoning, radiation side effects, schizophrenia, drug withdrawal, and Parkinson's disease.

d. Glutamic Acid

Glutamic acid play an important role in amino acid metabolism because of its role in transamination reactions and is necessary for the synthesis of key molecules, such as glutathione which are required for removal of highly toxic peroxides and the polyglutamate folate cofactors.

e. Glycine

Glycine plays an important role in metabolic regulation, preventing tissue injury, enhancing anti-antioxidant activity, promoting protein synthesis and wound healing, and improving immunity and the treatment of metabolic disorders like obesity, diabetes, cardiovascular diseases, ischemia-reperfusion, cancer, and various inflammatory diseases.

f. Tryptophan

Tryptophan is a precursor for serotonin, a brain neurotransmitter theorized to suppress pain. Free tryptophan enters the brain cells to form serotonin. Thus, tryptophan supplementation has been used to increase serotonin production in attempt to increase tolerance to pain. Tryptophan is also the precursor of melatonin, tryptamine, and kynurenine, and has an important role in the functioning of neurotransmitters like dopamine and nor-dopamine. Tryptophan supplement is used in treatment of pain, insomnia, depression, seasonal affective disorder, bulimia, premenstrual dysphoric disorder, attention deficit/hyperactivity disorder, and chronic fatigue.

g. Histidine

Histidine plays multiple roles in protein interaction and is also a precursor of histamine. It is also needed for growth and repair of tissue, for maintenance of the myelin sheaths, and in removing heavy metals from the body.

h. Lysine

Lysine is an EAA which is extensively required for optimal growth and its deficiency leads to immunodeficiency. Lysine is used for preventing and treating cold sores. It is taken by mouth or applied directly to the skin for this use.

i. Threonine

Threonine is used for treating various nervous system disorders including spinal spasticity, multiple sclerosis, familial spastic paraparesis, and amyotrophic lateral sclerosis.

j. Isoleucine

Isoleucine is a branched chain amino acid and is needed for muscle formation and proper growth. Chronic renal failure (CRF) patients on hemodialysis have low plasma level of the branched chain amino acids (BCAA) leucine, isoleucine, and valine. The abnormalities in the plasma amino acid pool can be corrected with appropriate high protein supplements.

k. Aspartic Acid

Aspartic acid (FAA) is the precursor of AAs methionine, threonine, isoleucine, and lysine and regulates the secretion of important hormones.

l. Serine

Serine is the precursor of glycine, cysteine, and tryptophan and plays many important roles in cell signaling. serine is also being used for treatment of schizophrenia.

4. Fish as Amino Acid Source

Fish is an important and cheaper source of quality animal proteins. Fishes seems to be unique among animals in having free histidine in its muscle. Histidine is found to be in higher amounts in dark meat fish such as dolphin, Japanese mackerel and Japanese pilchard (600-1300 mg percent of muscle), in fairly high amount in intermediate species such as barracuda, common seabass (10-400 mg percent) and in small amounts in white fish (5-48 mg percent).

Table 3.1: Fishes Rich in Particular Amino Acid

Amino Acids	*Species Recommended for Particular Amino Acid Deficiency*
Arginine	*Oncorhynchus mykiss, Tor putitora, Neolissochilus hexagonolepis*
Histidine	*Rastrelliger kanagurta, Catla catla, Stolephorus waltei, Amblypharyngodon mola, Puntius sophore*
Isoleucine	*Oncorhynchus mykiss, Labeo rohita, Stolephorus commersonii*
Leucine	*Stolephorus waitei, Rastrelliger kanagurta, Labeo rohita*
Lysine	*Stolephorus commersonii, Thunnus albacores, Tor putitora*
Methionine	*Stolephorus waitei, Tor putitora, Rastrelliger kanagurta*
Phenylalanine	*Cirrhinus mrigala, Catla catla, Labeo rohita*
Threonine	*Thunnus albacores, Nemipterus japonicus, Stolephorus waitei, Stolephorus commersonii*
Tyrosine	*Oncorhynchus mykiss, Tor putitora*
Valine	*Nemipterus japonicas, Cirrhinus mrigala, Rastrelliger kanagurta*
Tryptophan	*Tor putitora*
Glutamine	*Cirrhinus mrigala, Catla catla, Labeo rohita*
Glycine	*Cirrhinus mrigala, Catla catla, Labeo rohita*
Alanine	*Nemipterus japonicus, Labeo rohita, Catla catla*
Aspartic acid	*Stolephorus commersonii, Heteropneustes fossilis, Clarias batrachus*
Serine	*Stolephorus commersonii, Nemipterus japonicas, Thunnus albacares*

Table 3.2: Amino Acid Composition of Fish Protein (mg%)

Amino Acid	*Mackeral*	*Pomfret*	*Sardine*	*Oil Sardine*	*Hammer Head Shark*	*Silver Belly*	*sol*
Leucine	11	18.4	12.4	13.4	9.1	5.5	6.7
Phenylalanine	4.2	6.2	4.8	6.1	5.4	5.4	3.8
Valine	6	6.4	8.8	5.9	5.6	7.3	6.9
Methionine	4.3	4.7	2.9	3.8	3.1	2.1	1.9
Tryptophan	1.1	1.8	Traces	Traces	2.1	-	-
Threonine	5.5	6.4	4.9	4.7	4.7	5.8	5.6
Arginine	6.8	8.2	6.2	7	6.3	13.5	19.2
Lysine	6.5	3.7	6.1	8	5.1	6.4	7.4
Histidine	1.9	0.7	0.8	1.7	3.1	3.9	2.4

5. Shellfish as Amino Acid Source

- ☆ Crustaceans have higher content of free α-amino acids than that of fish
- ☆ Molluscs possess amino acid content with the range as that in between fish and crustaceans.
- ☆ Crustaceans and molluscs have small amounts of free histidine (5-30 mg percent).
- ☆ Crab (*Scylla serrata*) meat collected from Indian water was found to have 40.6 mg histidine per 100g. In crab body meat, predominant free amino acids are constituted by α-alanine and glycine
- ☆ Claw meat have 75.0 mg/100g of histidine. Lysine is predominant in claw meat.
- ☆ Free tyrosine and phenylalanine have been detected in almost all the species of prawns of Indian waters.
- ☆ High content of free alanine (460.8mg/100g wet muscle) and free glycine (1000mg/100g wet muscle) is found in *Penaeus semisulcatus.*
- ☆ High content of free arginine is found in *Metapenaeus dopsoni, Penaeus indicus* and *Penaeus monodon* (above 400 mg/100g wet muscle).
- ☆ Free proline content was high in *Metapenaeus dopsoni* and *Penaeus indicus* (above 1000mg/100g wet muscle).
- ☆ Among mollusk, *Crassostrea madrasensis* was found to have high arginine (2.7g/100g protein) and glutamine (2.5g/100g protein) followed by glycine (1.9 g/100g protein).
- ☆ Leucine, lysine and alanine were also found to be above 1g/100g of protein.
- ☆ Cysteine was found to be less.
- ☆ Tryptophan and aspartic acid were not found.
- ☆ *Perna viridis* was found to have all the amino acids below 1/100g of protein.

Tryptophan was not detected in this species.

Chapter 4

Food and Fish Lipids

1. Introduction

Fat is a generic term for a class of lipids. Fats are produced by organic processes in animals and plants. These are extracted and used as an ingredient. All fats are insoluble in water and have a density significantly below that of water (i.e. they float on water). Fats that are liquid at room temperature are often referred to as oil. Most fats composed primarily of triglycerides, some monoglycerides and diglycerides are also present. Products with a lot of saturated fats tend to be solid at room temperature. Products containing unsaturated fats, which include mono unsaturated fats and polyunsaturated fats, tend to be liquid at room temperature.

2. Types of Fat

2.1. Predominantly Saturated Fats (Solid at Room Temperature).

All animal fats (e.g. milk fat, lard, tallow), as well as palm oil, coconut oil, cocoa fat and hydrogenated vegetable oil are considered as predominantly saturated fats.

2.2. Predominantly Unsaturated and remain Liquid at Room Temperature

Vegetable fats -from olive, peanut (groundnut oil), maize (corn oil), cottonseed, sunflower, and soybean. However, both vegetable and animal fats contain saturated and unsaturated fats. Some oils (such as olive oil) contain in majority of mono unsaturated fats, while others with a high percentage of polyunsaturated fats (sunflower).

2.3. Saturated Fats

If the fatty acid has all the hydrogen atoms it can hold, it is said to be saturated. This type of fat is typically found in large amounts in foods/products from animals, e.g. meat, butter, cheese and cream. Many baked goods such as cakes, biscuits and pastries are also high in saturated fat. Excessive intake of saturated fat can increase blood cholesterol levels.

2.4. Unsaturated Fats

In unsaturated fats, some of the carbon atoms are joined to others by a double bond and therefore, could accept more hydrogen atoms. They are not completely saturated with hydrogen, so they are called as unsaturated fats.

There are two main types of unsaturated fats

1. Monounsaturated (containing one double bond) and

2. Polyunsaturated (containing more than one double bond).

Most monounsaturated and polyunsaturated fats have good qualities, with one exception -trans fatty acids. Trans fatty acids are, an unsaturated fat but offer no health benefits.

2.4.1. Monounsaturated Fatty Acids

There is one double bond present in its structure. Monounsaturated fatty acids do not raise blood cholesterol, and evidence shows that they reduce blood cholesterol levels if they replace saturated fat in the diet. Oleic acid is the main monounsaturated fat in our diets and this is sometimes called omega-9 (because the double bond is in position 9 of the fatty acid chain). Found in significant amounts in most types of nuts, avocado, pears, grape seed oil and olive oil and spreads made from these.

2.4.2. Polyunsaturated Fatty Acids

There is more than one double bond present in its structure. They come mostly from vegetable sources, such as sunflower oil or seeds, but are also found in, nuts, green leafy vegetables and oily fish such as mackerel and sardines. Polyunsaturated fatty acid can actively reduce blood cholesterol levels. The polyunsaturated fats found in oily fish specifically appear to have no effect on blood cholesterol levels, but they do alter the consistency of blood. There are two 'series' of polyunsaturated fats in food. They are also known as essential fatty acids. They are omega 3 and omega 6. Essential fatty acids are so called because the body cannot make them but they are essential to the body's normal functioning, therefore, must be supplied through diet. Humans are unable to make these essential fatty acids, because they do not have the particular desaturase enzymes that insert double bonds in position 3 and 6 of the fatty acid chain. Fats also carry the fat soluble vitamins A, D, E and K. Fats and lipids are energy storage materials in plants and animal tissues.

2.4.3. cis

A cis configuration means that two carbons are on the same side of the double bond.

2.4.4. trans

A trans configuration, by contrast, means that the next two carbon atoms are bound to opposite sides of the double bond. As a result, they don't cause the chain to bend much, and their shape is similar to straight saturated fatty acids. Most fatty acids in the trans configuration (trans fats) are not found in nature and are the result of human processing (*e.g.* hydrogenation).

The differences in geometry between the various types of unsaturated fatty acids, as well as between saturated and unsaturated fatty acids, play an important role is biological processes and in the construction of biological structures (such as cell members).

3. Fish Lipids

Lipid in fish generally carries natural flavour components. A certain amount of fat and fatty acid assists in providing smoothness of texture during mastication of lean fish. In fatty fish, the influence of fat on texture is even more important. The lipid content of fish varies widely from species to species and even within the same species from one individual to another depending on age, sex, environment and season.

Table 4.1: Lipid Content of Seafood

Types of fish	*Fat (%)*
Fatty fish	10
Lean fish	0.5
Crustaceans	2.1
Mollusks	1.5

Table 4.2: Variation of lipid content of dark and white meat of some fishes

Fish species	*Kind of meat*	*Crude fat%*
Tuna	Dark	6.7
	White	4.5
Sardine	Dark	12.8
	White	2.9
Mackerel	Dark	29.7
	White	13.1
Herring	Dark	28.2
	White	13.0

3.1. Distribution of Fat in Fish

The term lipid will be used for total fat component in fish. However, term fat is used for selected anatomical deposits, which are mostly triglyceride. In lean fish, the dark (red or lateral line) muscle has about twice the lipid of white muscle. The percentage of cellular lipid in the white muscle is normally altered by season. The lean fish generally have more fat in livers (e.g. cod) which show seasonal variation. In the fatty fish species, the muscle shows fluctuating levels of seasonal variation in neutral fat.

3.1.1. Types of fat

a. Neutral fat (Triglycerids)

b. Basic cellular lipids (Phospholipids)

c. Fatty acids

d. Wax esters

e. Glyceryl ethers

3.1.1.1. Neutral fat (Triglycerids)

It forms the major constituent of fish lipid. There are variations in the amount of neutral fat in muscle. The belly flap is a high fat section of many fishes, (eg. In mackerel -29% lipid in belly flaps, 18.3% lipid in dark muscle and 7.6% in while muscle). In male mackerel the skin fat forms 40% of the total fat in the whole fish. Triglyceride distributed through fish muscle tends to have a homogeneous fatty acid composition. Most species of marine organisms try to obtain an optimal fat and fatty acid composition, and their behavior and food preferences lead towards this objective.

3.1.1.2. Basic cellular lipids (Phospholipids)

Lean white fish muscle contain a minimum of about 0.7% of basic cellular lipid, 85-95% is 'polar' lipids, mostly phosphatidyl ethanolamine and phosphatidyl choline. The balance of this type of basic lipid includes sterol ester and free sterol, free fatty acids and triglyceride. This basic mixture represented the structural lipid of cell walls and that any excess of triglyceride and/or certain other non-polar lipids such as wax esters or glyceryl ethers, provided the 'fat' of fatty fish. Lipids are non-polar (hydrophobic) compounds, soluble in organic solvents. Most membrane lipids are amphipathic (having both hydrophilic and hydrophobic parts), having a non-polar end and a polar end. Fatty acids consist of a hydrocarbon chain with a carboxylic acid at one end."

3.1.1.3. Fatty Acid

The lipids extracted from edible parts or more specifically skeletal muscle of a fish is subjected to analyse for fatty acid profile or composition. This is ordinarily referred to as fatty acid profile of a fish.

- ☆ In a fatty or semi-fatty fish, fatty acids are derived mainly from neutral fat (triglycerides) and to a minor extent from phospholipids, cholesteryl esters and other esters.
- ☆ In a lean fish, the contribution made by phospholipids is substantial.

- ☆ Chain length of fatty acids reported for tropical Indian fishes and other aquatic animal foods ranges from C14 to C22, and is similar to those from non-tropical waters.
- ☆ C12 to C13 fatty acids in small or trace amounts have been reported in certain species and occasionally, all over the world.
- ☆ The special feature of lipids of fish and aquatic invertebrates both from tropical and non-tropical waters is the presence of substantial quantities of PUFA belonging to omega-3 series more particularly EPA and DHA.
- ☆ Equatorial Indian species appear to have more saturated fat compared to those from non-tropical fishes of Northern and Southern hemispheres
- ☆ Lipids of Indian species contain ordinarily lesser amounts of MUFA
- ☆ Species from non-tropical regions of Northern and Southern hemispheres contain in general more of EPA and lesser amount of DHA, the ratio of EPA to DHA is high (sometime as high as 30).
- ☆ Lipids from freshwater fish from subtropical and temperate regions are richer in C20 and C18 fatty acids and poorer in C20 and C22 fatty acids compared to marine fish.

Table 4.3: Comparative View of Omega 3 Fatty Acids in Fish

(more than 1.0 grams) per 3 Ounces	*(0.5 to 0.9 grams)*	*(less than 0.5 gram)*
	Scallops	Carp
Turbot	Crab	Cod
Salmon	Shrimp	Grouper
Herring mackerel		
Sardines	Sea bass	Pacific halibut
Atlantic bluefish	Snapper	Ocean perch
Most shellfish	Clams	Mahimahi
Pacific halibut	Lobster	Orange roughy
Squid	Lobster	
Anchovy	Striped bass	
	Shark	
	Mussels	
	Rainbow trout	

Table 4.4: Fatty Acids Distribution of Fish Oils

Fish	*Saturated*				*Unsaturated*					
	C14	*C16*	*C18*	*C20*	*C14*	*C16*	*C18*	*C20*	*C22*	*C24*
Thunnus thunnus	4.2	18.6	3.5	-	-	6.2	26.0	23.5	18.0	-
Stromateus cinerus	4.8	20.6	11.2	-	1.4	9.2	40.4	7.5	-	-
Stromateus niger	4.4	13.3	7.3	0.5	2.4	18.8	35.3	11.2	6.8	-
Hilsa ilisha	5.3	23.5	8.9	0.02	1.3	6.8	44.3	9.5	0.5	-
Clupea harengus	7	11.7	0.8	0.1	1.2	11.8	19.6	25.9	21.6	0.1 (2)
Trachurus trachurus	7.3	13.1	0.4	0.4	2.8	14.1	19.0	19.4	20.7	0.2 (2)

Table 4.5: Fatty Acid Composition of Phospholipids and Triglycerides or Non-phosphorylated Lipids of Oil Sardine, Indian Mackerel, Pomfret and Marine Catfish

Fatty Acid	*Oil Sardine (Wt percent of methyl ester)*		*Indian mackerel (Mole percent of methyl ester)*		*Pomfret (Percent by weight)*		*Marine Catfish*	
	PL	*TG*	*PL*	*NPL*	*PL*	*NPL*	*PL*	*NPL*
Saturated	17.5	25.8	19.3	24.3	31.3	28.9	16.2	28.5
16:0								
18:0	5.9	5.0	15.7	5.8	6.5	6.1	14.5	10.5
Others (C)	3.9	10.4	4.8	11.9	1.2	14.0	3.6	2.2
Monoenoic								
16:1	6.5	19.1	6.6	22.8	4.7	15.8	4.9	10.9
18:1	10.8	13.6	20.0	12.2	22.3	13.8	14.8	18.4
Others (d)	9.4	6.2	7.9	6.4	9.8	7.5	0.2	0.5
Total	26.7	38.9	31.5	41.4	36.8	37.1	19.9	29.8
Polyenoic	2.3	2.5	1.8	6.9	2.9	2.4	1.1	1.2
18:2								
18:3	1.7	0.5	0.7	1.2	2.8	0.7	3.2	2.5
20:5	10.6	7.9	5.0	8.6	5.2	3.2	5.8	4.8
22:6	28.0	4.6	17.5	Trace	14.7	4.7	18.6	10.6
Others (e)	3.9	4.8	0.2	—	0.2	2.8	17.1	10.0
Total	46.5	20.3	25.2	16.7	25.8	13.8	45.8	29.1

PL: Phospholipids; TG: Triglyceride; NPL: Non-Phosphorylated lipids

Table 4.6: Cholesterol Content (mg/100g muscle) of Fish

	Dorsal White Muscle	*Ventral White Muscle*	*Lateral Dark Muscle*
Juvenile	16	30	52
Medium Size	30	47	64
Large Size	46	79	85

4. Role of Fish Lipids in Human Nutrition

To achieve a balanced diet, there is a need to reduce total fat intake and it is also important to make sure that the type of fat we eat is right. The fat that is not beneficial to human is the hard "saturated" fat which comes mainly from the fat of land animals such as cows and sheep. Fish fat has several beneficial effects.

4.1. Provide Food with Low Fat

Fish is a good food for a low fat diet. It is low in calories and many types of fish do not contain many saturated fat. The nutritional value of fish will vary slightly according to the location it is also harvested, the cut of fish, and the age of the fish. The method used for cooking will have an effect on it.

4.2. Reduce Cholesterol Level in Blood

Cholesterol is a type of fat which is naturally produced by our body and is also found in the diet. It is usually deposited in the lining of blood vessels, causing them to narrow. Then the heart has to work harder to pump blood around the body. Cholesterol deposits can cause blood clotting. Tissues will be deprived of oxygen when the blood vessels are blocked. Unsaturated fats can help to reduce the cholesterol level in the blood, thus lowering the risk of heart disease. Oil-rich fish such as mackerel, sardines, herring and sprats are rich in unsaturated fats containing Omega-3 that is valuable for health.

4.3. Health Benefit of Omega-3 Fatty Acids

Two fatty acids, eicosapentaenoic acid (EPA) and docosahexaenoic (DHA), collectively known as Omega-3, are essential fatty acids. Schizophrenia symptoms can be eliminated or at least vastly diminished by oral supplementation with EPA. DHA is the building block of human brain tissue and is particularly abundant in the grey matter of the brain and the retina. Low levels of DHA have been associated with depression, memory loss, dementia and visual problems. DHA is particularly important for fetuses and infants; the DHA contents of the infant's brain triples during the first three months of development. Optimal levels of DHA are therefore crucial for pregnant and lactating mothers. Oil-rich fish, such as salmon, trout, mackerel, herring and sardines, are an excellent source of Omega-3 fatty acids, which are essential in our diet. Eating oil-rich fish provides the Omega-3 fatty acids needed for the body. Omega-3 oils from fish have a lowering effect on blood fats. This decreases the chance of the blood vessels clogging with cholesterol and reduces the risk of dying from heart attacks. Omega-3 can also make blood less "sticky", and it therefore flows more easily around the body. This can reduce the risk of a heart attack. They also help to reduce blood pressure to some extent and keep the heart beat steady.

4.4. Fish Oils Prevent Cancer

Fish oils can help to prevent cancer cells progressing to the tumor stage. They may also reduce inflammation and provide relief for people suffering from rheumatoid arthritis and even some skin disorders such as psoriasis.

4.5. Needed for the Development of Brain

Omega-3 oils can play an important part in aiding the development of brain. Expectant mothers are advised to eat a lot of oil-rich fish in the last three months of pregnancy to assist the baby's brain growth. A good supply of Omega-3 oils assists the development of nerves and eyesight.

Chapter 5

Extractive Components of Fish

1. Introduction

Compounds of extractive fraction are present in plasma and intercellular fluid of the cells and are readily extractable from tissues when treated with trichloroacetic acid or ethanol to precipitate proteins. These compounds are more vulnerable to the action of bacteria, so their content and nature affect the spoilage pattern of fish and shellfish. They provide characteristics odour and flavour of a species.

2. Non-Protein Nitrogenous Compounds (NPN)

Nitrogen containing extractives are defined as the water-soluble, low molecular weight, nitrogen containing compounds of non-protein nature. This NPN fraction constitutes from 9-18 percent of the total nitrogen in teleosts, about 25 percent in crustaceans and molluscs and about 30 percent in elasmobranchs. Muscles of elasmobranch fishes are unique in having high amounts of non-protein nitrogen and urea. Non-protein content in elasmobranch is predominatly contributed by urea and to a lesser extent by trimethylamineoxide. Compared to marine teleosts, fresh water fish have very low (almost insignificant) concentration of trimethylamine oxide in their flesh. High levels of free α-amino acid nitrogen are present in the muscle of crustaceans and molluscs (about 300 mg per 100g) whereas flesh of teleosts has much lesser amounts (about 30mg per 100g wet flesh for marine species and about 50mg per 100g for fresh water fish). Elasmobranchs appear to have higher amount free α amino acid nitrogen content than teleosts (about 75mg).

Table 5.1: NPN Components in Seafoods

Compound (mg/100g)	*Cod*	*Herring*	*Shark*	*Lobster*
Free amino acids	75	300	100	3000
Creatine	400	400	300	0
Betaine	0	0	150	100
TMAO	350	250	500-1000	100
Anserine	150	0	0	0
Urea	0	0	2000	-

It comprises of

1. Tri methyl amine oxide (TMAO)
2. Urea
3. Alpha amino nitrogen(AAN)
4. Free amino acids,
5. Low molecular weight peptides
6. Nucleotides and related compounds
7. Organic bases

2.1. Free Amino Acids

High content of histidine is noted in active migratory fishes such as tuna and skipjack. These fishes are red-fleshed, but some white-fleshed freshwater fishes *e.g.* carp and crucian carp also contain high histidine. High taurine is noted in white-fleshed marine and fresh water fishes.

Crustaceans

Free amino acid is high in crustaceans when compared with that in fish. High levels of taurine, proline, glycine, alanine and arginine found in crustaceans. Prawns contain high concentration of glycine *i.e.* 1 percent of fresh muscle than crab. Glycine is therefore related to the sweetness of prawn muscle. Crab tends to accumulate more taurine.

Mollusks

Mollusks lie between fishes and crustaceans in terms of free amino acids content. Mollusks are rich in taurine, proline, glycine, alanine and arginine.

2.2. Nucelotides and Related Compounds

Nucelotides serve as important palatable taste (umami) producing factors. IMP(Inosine mono phosphate) and GMP (Guanosine mono phospahte) show a distinct taste enhancing effect in combination with glutamic acid. About 96 percent of nucleotides are accounted by purine derivatives and small amounts by uracil and cytosine derivatives. In the live muscle, ATP predominates under normal conditions. After death, it is enzymatically degraded by the following pathway in fish:

ATP → ADP → AMP → IMP → Inosine → Hydroxanthine.

The reaction, IMP → Inosine is rather slow, hence IMP accumulate in fresh fish muscle. In crustaceans, AMP tends to accumulate because of low AMP deaminase activity. Mollusks rarely have AMP deaminase and the major pathway of ATP degradation is: ATP → ADP → AMP → Adenosine → Insoine → hypoxanthine.

Presence of small amounts of di-and tri-phosphopyridine nucleotide (DPN, TPN), guanosine triphosphate (GTP) and uridine triphosphate (UTP) and traces of guanosine 5-momo- and diphosphates (GMP,GDP), uridine 5-mono and diphosphates(UMP, UDP), UDP- glucose, UDP-acetylglucosamine and UDP-glucaronic acid were also noted. In leg meat of boiled carb, AMP was detected as a main component in addition to small amounts of ADP, IMP, and UMP.

2.3. Guanidino Compounds

Creatinine and arginine predominate in fish and invertebrates. Creatinine, which is formed through dehydration of creatine is present in smaller amounts. Octopine is guanidino compound peculiar to mollusks, making up 10-20 percent of the total extractive nitrogen in cephalopods. Octapine increases with a decrease of arginine phosphate in scallop, squid and octopus.

2.4. Urea

Urea in muscle of teleosts is usually below 50mg. In elasmobranchs, it amounts to 1400-2000 mg and is the most prominent nitrogen component, serving not only for detoxification of ammonia but also for osmoregulation.

2.5. Quarternary Ammonium Bases

Common bases found in fish/shellfish are trimethylamine oxide (TMAO) and betaines. The TMAO levels in marine teleosts fluctuate greatly depending on season, size of fish and environmental conditions. TMAO has not been detected in fresh water teleosts. Among invertebrates, squid appear to have high TMAO (100 – 1000 mg). TMA is known as principal producer of fish odour and is scarely detected in fresh muscle. It increase gradually with the post-mortem bacterial reduction of TMAO.

Betaines are minor components in fish muscle extracts but are major compound is extracts of crustaceans and mollusks. The main compound is glycine betain and their levels range from 400 -900 mg in crustaceans and mollusks. Homarine is another compound widely distributed among marine invertebrates. The occurance of other betaines such as γ butyrobetaine, carnitine, trigonelline and stachydrine are present in small quantities.

3. Other Nitrogenous Compounds

Arginine has been isolated from the fan mussel, α-amino propioacetic acid from scallop and β-hydroxyaspartic acid from shortneck clam. Alanine betaine (β-homobetaine) was found in fan mussel, oyster, scallop and krill.

4. Miscellaneous Constituents

4.1. Bile Salts

Bile salts which synthesised in the liver from cholesterol are extremely bitter. In a few species of fish, particularly Indian major carps of fresh water habitat, gall bladders which contain lower level bile salts (when expressed as the percentage of total body weight) compared to that of a marine species.

During storage in ice or at ambient temperature or during bad handling and filleting particularly when the fish is at its lowest limit of edibility, gallbladder may get raptured and bitter bile makes the ventral belly portion unfit for consumption. *Labeo rohita* has the highest amount of bile content in relation to its body weight and their species is more susceptible to give above trouble. Marine species such as oil sardine, Indian mackerel, seer fish or tuna have too thin gallbladder and bile of those fish constitute a negligible fraction of the body weight.

4.2. Squalene

Liver oil of certain species belonging to the family of shark has been shown to contain a highly unsaturated hydrocarbon, squalene ($C_{30}H_{50}$)-dihydrotriterpene with six double bonds per mol. In a few other species of fish also, squalene has been found in small amount. Several other hydrocarbons have also been identified in liver oils of certain species of shark and other elasmobranch fish. These are saturated pristane ($C_{18}H_{38}$) and monounsaturated zamene ($C_{18}H_{36}$ Liver oils of certain species of sharks, especially the deep sea sharks, may contain as high as 80 percent unsaponifiable matter of which 90 percent may be squalene and 7-45 percent of liver oil of basking shark (*Cetorhinus maximus*) is composed of squalene. Squalene and its companions have also been found in shark eggs and stomach lipids.

In general, squalene is found chiefly in fish liver oils with high content of unsaponifiable matters. Of the liver oils from four species of sharks of tropical sea of India, only the oil from *Chiloscyllium* had 80 percent unsaponifiable matters, of which 71 percent was squalene.

Squalene is an intermediate compound in the biosynthesis of cholesterol. Its use in the manufacture of pharmaceuticals, aromatics, surface active agents, rubber chemicals and also as an additive to cosmetics, has been reported. The compound has commercial importance.

4.3. Alkoxydiglycerides

Among elasmobranch fishes, alkoxydiglycerides are of common occurrence in lipid of visceral depot and also in liver oils. All such fishes have large oil-rich livers. In many elasmobranchs, the liver depot lipids consist only of triglycerides e.g. thresher shark, skate. In *Squalus acanthias,* a mixture of alkoxydiglycerides and triglycerides is found. At the same time, depot lipids of rat fish may contain 30 to almost 100 percent alkoxydiglycerides with a squalene content ranging from virtually nil to about 60 percent.

4.4. Wax Esters

Wax esters form minor components of fish liver lipids. They occur in large proportions, in the liver lipids of two Japanese species.

(Lotella phycis and *Aaemonema morosum)* Wax may also form an important fraction of the total lipids of the ovaries of some fish. Dried and salted ovaries of the grey mullet (*Mugil japonicus*) is a highly esteemed food in Japan. The product as eaten contains around 25-30 percent lipids, of which nearly 90 percent is wax esters. Total lipid of roe of mullet (*Mugil parsia*) has been reported to be about 8.4 percent ; of this, 67 percent is wax ester, 18 percent phospholipids and only about 8 percent is neutral lipid. Fatty acid and alcohol composition of wax ester isolated from the roe lipid of *M. parsia* indicate less unsaturated fatty acid and mostly long chain saturated alcohols.

Chapter 6

Vitamins in Fish

1. Introduction

Vitamin are vital substances, needed in very small quantity for growth and good health. Vitamins are key regulators in the body. Most of them are not synthesized by the body and must be supplied through diet, except a few whose requirement may be partially met by synthesis in the body (*e.g.* folic acid and vitamin D).

There are 13 essential vitamins: Vitamins A, C, D, E, K and B vitamins. The B-complex family of vitamins contains 8 of the 13 essential vitamins needed for good health.

1. Vitamin B-1: Thiamine
2. Vitamin B-2: Riboflavin
3. Vitamin B-3: Niacin
4. Vitamin B-5: Pantothenic Acid
5. Vitamin B-7: Biotin
6. Vitamin B-6: Pyridoxine
7. Vitamin B-9: Folate
8. Vitamin B-12: Cobalamin

2. Classification

Vitamins are classified according to their solubility in fat and water.

2.1. Fat Soluble Vitamin

1. Vitamin A or Retinol
2. Vitamin D or Cholecalciferol

3. Vitamin E or Tocopherols
4. Vitamin K or Naphthoquinones

2.2. Water Soluble Vitamin

1.Thiamin or vitamin B1
2. Riboflavin or vitamin B2
3. Niacin (nicotinic acid and nicotinamide) or vitamin B3
4. Pyridoxine or vitamin B6
5. Pantothenic acid or Vitamin B5
6. Biotin or Vitamin B7
7. Cyanocobalamin or vitamin B12
8. Folate (folic acid) or vitamin B9
9. Ascorbic acid or vitamin C

3. Vitamin content of Edible Flesh of Fish

- ☆ Fish are good source of fat soluble vitamins A,D and E and of water soluble vitamins of B group.
- ☆ Fish liver oils are known as the richest source of vitamin A and D. Vitamin A upto 30,000 IU/g and vitamin D upto 250,000 IU/g of liver has been found in some species.
- ☆ Fish liver oils are often used for medicinal purposes because of these vitamins.
- ☆ Fish oils are also rich in vitamin E than any other animal fat.
- ☆ Fish flesh is an excellent source of thiamine, riboflavin, niacin, pyridoxine, biotin, pantothenic acid and vitamin B12.
- ☆ Shellfishes are also a rich source of vitamin B.

Table 6.1: Vitamin Content of Edible Flesh of Fish

Vitamin	*Average Content (μg percent)*	*Vitamin*	*Average Content (μg percent)*
Vitamin A	25	Pantothenic acid	500
Thiamine	20	Pyridoxine	500
Riboflavin	120	Biotin	5
Nicotinic acid	3000	Folic acid	80
Vitamin B^{12}	1	Vitamin C	3000
Vitamin E	12	Vitamin D	15

4. Vitamins and its Significance

Vitamins are classified according to their solubility in fat and water.

4.1. Fat Soluble Vitamins

- ☆ Fish is not ordinarily recommended as a major source of vitamin A, vitamin D or vitamin E.
- ☆ Most of the species of fish are poor suppliers of these vitamins.
- ☆ Vitamin A content of flesh has been reported to vary from 0 to 300 I.U/ and vitamin D content from 0 to 17 I.U/g flesh or 20 to 160 I.U/g body/ flesh oil.
- ☆ Flesh of fresh water eel, dogfish, lantern fish, lizard fish and lamprey are particularly rich in vitamin A content (500-9800 I.U/g tissue); flesh of eel is rich in vitamin D (4700 I.U./100g).
- ☆ Liver oils of many species of fish are rich in vitamins A and D. Leading sources are livers of cod, halibut and tuna. Vitamin D content of liver oils of above fishes were found to vary from 550 I.U to 2,50,000 I.U. per g. Liver oils of cod and few other species of fish have been reported to contain vitamin E (about 30mg/100g oil for cod and 18 to 45 mg/100g oil for others).
- ☆ Liver oil of sharks which are abundantly harvested from seawaters of India was found to be rich in vitamin A and comparable to livers oil of cod and halibut of temperate regions.
- ☆ Vitamin A content of shark liver oil of Indian subcontinent ranges from 4700 to 32900 I.U/g. The oil is poor in vitamin D (6.3 to 220 I.U/g).
- ☆ Based on these investigations, production of shark liver oil as a source of vitamin A for human consumption on a commercial basis started in many maritime states of India.
- ☆ Liver oils of both freshwater and marine fish of India contain vitamin A1, vitamin A2 and neo-vitamin A1(geometrical isomer of Vit. A).
- ☆ Liver oils of freshwater species consist predominantly of vitamin A2 (vitamin A2 : vitamin A1 = 9:1). Vitamin A1 continues the predominant fraction of shark liver oil.
- ☆ The vitamin D content of liver and body oils of freshwater fish of Bengal is very poor. The value for vitamin D content of shark liver oil ranges from 150 to 260 I.U. with an average of 200 I.U. per g. In a fish, liver oil contains the highest amount of vitamin E. Liver oil of shark of Indian sea was found to have 1 to 2 mg/g of oil.

4.1.1. Vitamin A or Retinol

Vitamin A or retinol is a pale yellow solid which dissolves freely in oils and fats but is only very slightly soluble in water. It is found in the fatty parts of foods, for example, in the fat of milk, butter and fish liver oils. A small amount is present in green vegetables and carrots. Vegetables contain no retinol as such, but pigments called carotenes, which are chemically related to it. Carotenes can be converted to retinol in the wall of the small intestine during absorption and hence vegetables have considerable vitamin A activity. Several carotenes are known, but the most important is β-carotene, which is often referred to as provitamin A.

Functions

Vitamin A is mainly associated with vision and tissue growth.

Deficiency

A long-term deficiency of vitamin A may lead to a condition known as 'night blindness' which makes difficult to see in a dim light. Long-term deficiency may cause an eye disease known as xerophthalmia in which dead cells accumulate on the surface of the eyes causing them to become dry and opaque. The cornea may become ulcerated and infected-a condition known as keratomalacia-and blindness is a common sequel.

RDA (Recommended Daily Allowances)

RDAs for vitamin A are expressed in terms of retinol activity equivalents (RAE). The RDA for adult men and women is 900 and 700 μg RAE, respectively. The requirement is same for pregnant and lactating women.

Sources

Retinol is found in animal tissues (especially liver) and dairy products. Fish-liver oils are the richest source. Carotenes are found in plant tissues. Carrots, dark green vegetables and yellow fruits are good sources of carotenes. Spinach, with its dark green leaves, is well endowed with carotenes. Milk and milk products are also good sources of vitamin A.

4.1.2. Vitamin D or Cholecalciferol

Naturally occurring vitamin D is more frequently referred to as vitamin D_3 or cholecalciferol. Another form of vitamin D, known as vitamin D_2 or ergocalciferol, can be made by exposing the compound ergosterol, which is found in fungi and yeasts, to ultraviolet light. Cholecalciferol, vitamin D_3, is the only one of these compounds of dietary importance. Vitamin D is measured in terms of international units (IU) of cholecalciferol (Vitamin D_3). One IU is equivalent to 0.025 mg of pure crystalline vitamin D_3 or the biological activity of 40 IU is equal to 1 μg.

Functions

Vitamin D is needed for the absorption of calcium and phosphorus from digested food. In its absence, the body is unable to make use of these elements and they are lost in the faeces. Phosphorus and calcium are both needed to form bones.

Deficiency

Deficiency of vitamin D causes rickets in the young and related bone disease, osteomalacia (weaking the bones). Rickets is characterized by curvature of the bones in the limbs and other symptoms of improper bone formation caused by loss of calcium from the bones and its replacement by softer tissue.

RDA

400 to 800 IU per kg body weight per day. The RDA for adult men and women is 600 IU which is also same for pregnant and lactating women.

Sources

Egg yolks, are the best source, of vitamin D. The content vary depends mostly upon the content of the vitamin in the hen's diet. Diary products contain some vitamin D, the potency varying with the season. Livers of fish and oils extracted from the livers are extremely rich sources.

4.1.3. Vitamin E or Tocopherols

Vitamin E is the name given to α-tocopherol ($C_{29}H_{50}O_2$) and a group of fat-soluble saturated and unsaturated alcohols closely related to it.

Function

It performs a similar function in the body where it is major lipid-soluble antioxidant. It scavenges free radicals and prevents cell membranes and their contents from free radical indicated damage. Vitamin E also protects nutrients such as PUFAs, vitamin A vitamin C from oxidation. It also showed protective effect against diseases, such as some types of cancer, arthritis and ischaemic heart disease, in which free radicals are found to be involved. Vitamin E measured in terms of mg (milligram of D-alpha tocopheral acetate, or as tocopherol equivalent (TE).

Deficiency

Vitamin E deficiency disease has not been recognised. In severe deficiency, there is increased hemolysis and muscular dystrophy.

Daily Requirement

The dietary requirement of vitamin E is related to the PUFA content of the diet, and it is estimated that 0.4 mg of the vitamin is required for each gram of PUFA. RDA for vitamin E ranges from 4mg to 15mg.

Sources

Vitamin E is ubiquitous in its distribution and is found in vegetable fats, oils, dairy products, meat, eggs, cereals, nuts leafy green vegetables and yellow vegetables.

4.1.4. Vitamin K or Naphthoquinones

Vitamin K comprises several closely related fat-soluble compounds derived from menadione (2-methyl-1,4-naphthoquinone) all of which display vitamin K activity.

Functions

Prothrombinogen and other blood cotting factors are formed in the liver with the help of vitamin K. These factors are necessary for the clotting of blood.

Deficiency

Its deficiency interferes with the formation of prothrombinogen and thus reduces the clotting tendency of blood. In other words, the blood clotting time is prolonged. Clinically this condition is called hypoprothrombinemia. Internal or external haemorrhages may ensue either spontaneously or following injury or surgery. A premature infant is sensitive to lack of this vitamin.

Sources

Vitamin K is present in most foods, but green leafy vegetables are the richest source. Bacterial synthesis in the bowel provides humans with vitamin K in addition to that obtained from foodstuffs. In most cases, the amount made available in this way is sufficient to supply the body's requirements.

RDA

There is little danger of vitamin K deficiency in a person who eats a normal diet and, while there are no specified dietary reference values, a daily intake of 1mg per kilogram of body weight is thought to be both safe and adequate for adults. Newborn infants lack the bacteria which produce vitamin K in the gut and many are given supplements to prevent deficiency and the possible occurrence of life-threatening haemorrhagic disease. RDA for vitamin k ranges fom 2 μg to 120 μg.

4.2. Water-Soluble Vitamins

The water-soluble vitamins include members of the vitamin B complex and vitamin C. The B group of vitamins comprises several vitamins which have similar functions and are often found together in foods. In the body, they are largely concerned with the release of energy from food. They are all soluble, to a greater. Since the body lacks the capacity for storing them, any excess over immediate requirements is excreted in the urine.

The members of B- complex vitamins namely thiamine (B_1), riboflavin (B_2), niacin (Pellagra-Preventing factor), nicotinic acid and niacinamide, Pyridoxine, Pantothenic acid, folic acid, cyanocobalamin (B_{12}, antipernicious anemia factor, cobalamin), carnitine and choline are present in the muscle of fish and shellfish. In addition, water soluble vitamin C (ascorbic acid) is also present in fish flesh.

4.2.1. Thiamin or Vitamin B1

This is a white, water-soluble crystalline solid. The thiamine molecule has a complex structure which includes an amino ($-NH_2$) group and a hydroxyl group. Thiaminis esterified with pyrophosphoric acid in the body to give thiamine pyrophosphate, or cocarboxylase.

Function

The main function of B_1 is that act as a controlling agent in energy metabolism. It does so by acting as a coenzyme TPP in key reactions that produce energy from glucose or that convert glucose to fat for tissue storage. Thiamine pyrophosphate is also a coenzyme for transketolase, which is an enzyme that is required to produce keto acids.

Deficiency

Polyneurities or beriberi is a deficiency disease of thiamine. It causes loss of appetite and other symptoms such as irritability, fatigue and dizziness. Dry beriberi is characterized primarily by emaciation and multiple neurotic symptoms which proceed from foot, then to calf muscles and then to the thigh. Wet beriberi is characterized by a severe edema which masks the emaciation, enlargement of heart, tachycardia, dyspnoea and palpitations of the heart even on the slightest exertion. The nervous system is badly affected and this may produce partial paralysis and muscular weakness.

Sources

Thiamin plays an essential part in the utilization of carbohydrates by living cells. It is present in all natural foods to some extent. Unfortunately, it is often absent from processed foods because it has been removed or destroyed in the preparation of the food for the market. Polished rice and low extraction rate flour from which thiamin has been largely removed. Sugar, refined oils and fats and alcoholic beverages are examples of foods which contain little or no thiamine.

RDA

Requirment of thiamine is closely related to the calorific needs and increase in intake of calories also increases the thiamine requirement. The average adult requires from 0.23 to 0.5 mg per 1000 calories. The RDA is 1.2 mg. The RDA for pregnant and lactating women is 1.4mg.

4.2.2. Riboflavin or Vitamin B2

This is a yellowish-flurescent solid, like thiamine, it has a complex chemical structure.

Functions

In the body riboflavin is esterified with phosphoric acid or pyrophosphoric acid and froms part of two coenzymes (FAD and FMN) involved in a variety of oxidation. reduction reaction processes concerned with the release of energy form proteins, fats and carbohydrates in living cells. It controls removal of amino nitrogen from certain D and L amino acids.

Deficiency

A deficiency of riboflavin produces a check in the growth of children, lesions on the lips (cheilosis) and scaliness at the corners of the mouth (Angular stomatitis). The tongue and eyes may also become irritated (Glossitis).

RDA

The RDA for adult males is 1.3 mg per day, and for adult females 1.1 mg per day. During pregnancy and lactation the RDA is increased by 0.3 mg and 0.5 mg respectively.

Sources

Riboflavin is widely distributed in plant and animal tissues. The main sources of riboflavin are milk, cheese, meat, fortified breakfast cereals and eggs. Riboflavin is only very slightly soluble in water and losses by solution during cooking are small. Heating causes little breakdown of riboflavin and little or no loss occurs during canning. Meat losses about one-quarter of its riboflavin during roasting. Greater losses occur if riboflavin in heated under alkaline conditions like sodium bicarbonate is added to the water used for boiling vegetables.

4.2.3. Niacin (Nicotinic Acid and Nicotinamide)

The B group vitamin known as niacin exists in two forms: a pyridine carboxylic acid called Nicotinic acid, and its amide, nicotinamide. Unlike most other members of the B group of vitamins, these two substances have simple chemical structures. The acid and the amide are equally effective as vitamins.

Functions

Nicotinamide occurs in the body as part of two essential coenzymes, NAD and NADP, concerned in a large number of oxidation processes involved in the utilization of carbohydrates, fats and proteins. These coenzymes are also needed in the system that oxidizes glucose to release controlled energy.

Deficiency

A severe deficiency of niacin can cause the disease 'pellagra' which is characterized by dermatitis, diarrhoea and symptoms of mental disorder. Less severe deficiencies can produce one (or) more of these symptoms.

RDA

The amount of niacin required for the maintenance of good health is related to the energy content of the diet and to the amount of tryptophan present. The RDA for adult men is 16mg/day of niacin equivalents and for women is 14mg/day. Thus, for an adult man with a diet of energy content 2550 Kcal, the per day. The RDA is increased by 3 mg per day for lactating women. Additional amounts (4mg) are required by pregnant women though they are able to convert tryptophan to niacin about twice as efficiently as usual.

Sources of Niacin

Niacin is found in both animal and vegetable tissues. The main sources of the vitamin are meat and meat products, fish, pean and butter. Potatoes, bread, fortified breakfast and cereals are poor sources. Milk and eggs contain little niacin but their proteins are especially rich in tryptophan and so they serve as food sources of the vitamin.

4.2.4. Pyridoxine or Vitamin B6

Vitamin B_6 is the name given to a group of three pyridine derivatives, Pyrioxine, pyridoxal, pyridoxamine. All three compounds are interconvertable in the body and they are equally potent as vitamins.

Functions

Vitamin B_6 functions as a coenzyme for a large number enzymes involved in amino acid metabolism, and hence requirements are related to dietary protein intake. Deamination and transamination of amino acids to form new amino acids. This is important in the formation of amino acids to form new body. Decarboxylation of amino acids forms amines, which are active metabolic control agents. Thus serotonin, a potent constrictor of blood vessels which stimulates cerebral activity and brain metabolism, is formed from tryptophan. Trans-sulphuration of sulphur transfer reactions which help to form sulphur-containing amino acids from precursors, *e.g.* cystine is formed from methionine.

Deficiency

Symptoms of vitamin B_6 deficiency in animals can be produced by feeding them with a diet devoid of this vitamin. It is not easy, however, to do the same thing with humans, various skin lesions are reputed to be caused by vitamin B_6 deficiency. Infants fed on milk powders deficient in vitamin B_6 and were found to suffer from convulsions but responded rapidly to treatment with the vitamin.

RDA

The RDA for adult males is 1.3 mg per day, and for adult females 1.1 mg per day. During pregnancy and lactation the RDA is increased by 0.6 mg and 0.7 mg respectively.

Sources

Vitamin B_6 is found in foods which contain the other B vitamins. The main sources in the diet are potatoes and other vegetables, milk and meat.

4.2.5. Pantothenic Acid

This vitamin is a pale yellow oil. It is found in a wide variety of plant and animal tissues.

Functions

Pantothenic acid is of the highest biological importance because of its incorporation in coenzyme A. coenzyme A is the functional, active form of pantothenic acid in the body. It is involved in several vital enzymatic reactions. The most important of these reactions is the formation of active acetate which is an important common molecule in the process of energy production in the cell from carbohydrates and fats. It is also a precursor of cholesterol and steroid hormones. It is required for the production of haeme in the synthesis of haemoglobin.

RDA

It is so widely distributed that a normal diet, which contains 10-20 mg, should be adequate and there is no danger of its deficiency. The adequate intake (AI) level for adult men and women is 5mg.

Sources

Yeast, liver and kidney are rich sources. Good source include egg yolk and skimmed milk powder. Fair sources are lean meat, beef, cheese, legumes, sweet potatoes and yellow corn.

4.2.6. Biotin

Biotin is another widely distributed vitamin which is required in minute amounts.

Functions

It functions as a coenzyme cocarboxylase and involved in the metabolism of fats and carbohydrates.

RDA

Small amounts of biotin are required to the body and may be produced by the microorganisms present in the large intestine. The adequate intake (AI) level for adult men and women is 30 μg. The AI for pregnant and lactating women are 30 and 35 μg respectively.

Sources

Many foods contain biotin. Liver and kidney are good dietary sources and smaller amounts are found in egg-yolk, milk and bananas. Raw egg white contains a proper or protein-like substance called avidin which combines with the biotin of the yolk to form a stable compound. This is not absorbed from the intestinal tract and so the brain is not available to the body. Avidin can also render unavailable biotin in other foods. This activity is destroyed in cooking.

4.2.7. Cyanocobalamin or Vitamin B12

Cyanocobalamin is a deep red crystalline substance with the molecular formula $C_{63} H_{90} O_{14}$. It has by far the most complex chemical structure of any vitamin. The presence of a cobalt atom in the molecule is a noteworthy feature.

Functions

Cyanocobalamin plays an important role in the production of nucleic acid and in the complex process of cell division in the body. Vitamin B_{12} acts as an important coenzyme in the synthesis of deoxyribonucleic acid and vital proteins in the cell.

Vitamin B_{12} is necessary for the formation of red blood cells. It is also required for the activation of folic acid coenzymes.

Deficiency

Deficiency of vitamin B_{12} results in pernicious anaemia, in which extreme anaemia is accompanied by generation of the nerve tracts in the spinal cord. Pernicious anemia is caused by the absence of an intrinsic factor from the gut.

RDA

The amount of cyanocobalamin required for the maintenance of good health is exceedingly small and the RDA for those over 15 years of age is 2.4 μg. The RDA for pregnant and lactating women are 2.6 μg and 2.8 μg respectively.

Sources

Cyanobalamin found in small amounts in all animal tissues but it is absent from vegetable origin. Liver, kidney and lean meat are good sources. Eggs, milk and cheese supply a fair amount.

4.2.8. Folate

Folate is the name given to a group of closely related compounds derived from folic acid (pteroylglutamic acid).

Functions

Coenzymes of folic acid are known as tetrahydrofolates that are necessary for the interconversions of various amino acids and for the synthesis purine and pyrimidine. Folic acid to form haeme, the iron containing part of haemoglobin.

Deficiency

A deficiency of folate may cause megaloblastic anaemia. This is similar to the anaemia caused by nonabsorption of cobalamin but it is not accompanied by degeneration of the nerve cells which is feature of pernicious anaemia. Pregnant women are prone to develop this type of anaemia. Folate deficiency during pregnancy may be lead to fresh premature birth and low birthweight.

RDA

The RDA for folate for those over 19 years of age is 400 mcg dietary folate equivalents. The RDA level for pregnant and lactating women are 600mcg DFE and 500mcg DFE respectively.

Source

Good sources are fresh green leafy vegetables like spinach, lettuce, liver, kidney, dry beans and pulses.

4.2.9. Ascorbic Acid or Vitamin C

Ascorbic acid, or vitamin C is a white-soluble solid with the formula, $C_6H_8O_6$.

Functions

Vitamin C is required to build and maintain bone matrix, cartilage, dentine (tooth), collagen, capillaries and general body tissue. It is closely associated with proteins involved in tissue growth, tissue building and rebuilding, and with cell metabolic processes. It involves in the development of red blood cells by influencing the absorption of iron for haemoglobin formation. Healing of wounds including internal fractures of bones. Vitamins C is needed in large amounts for extensive tissue building during severe burns. It is also essential for healing bone fractures. Pregnancy, infancy and childhood demand more ascorbic acid. Vitamin C is necessary for the metabolism of tyrosine.

Deficiency

The cells of the body concerned in the formation of bone and the enamel and dentine of teeth lose their normal functional activity in the absence of ascorbic acid. A lack of ascorbic acid in the diet causes a condition known as 'scurvy', which is characterized by haemorrhages under the skin and in other tissues; and swollen and spongy gums from which the teeth are easily dislodged or may fall out. Scurvy infants is associated with great tenderness and pain in the lower limbs together with changes in the bones structure which are not found in adult scurvy.

Sources

Ascorbic acid occurs mainly in foods of plant origin. Green vegetables potatoes and fruit are the most important sources of ascorbic acid. Most fruits are good sources but apples, pears and plums supply negligible amounts.

Loss of Ascorbic Acid

Ascorbic acid in one of the most important nutrients, and it is easily destroyed by oxidation, exposure to light or high temperatures, alkalinity and metal ions. In extracts, juices and foods with cut surfaces, it may be oxidized at room temperatures by exposure to air. The oxidation is enzymically catalysed by oxidases which are contained within the cells of foodstuffs and are set free on cutting or crushing. Losses of ascorbic acid occur during storage of fruits and vegetables. Some less also occurs during the preparation and cooking of foods. This is due partly to oxidation and partly to solution in the water used for cooking.

RDA

For those over 15 years of age the RDA is 75 mg per day; smaller amounts are specified for younger age group, and larger amounts for pregnant and lactating women (80-120mg).

Chapter 7

Minerals in Fish

1. Introduction

Minerals constitute 3 – 4 percent of the total body weight. Mineral are inorganic elements occurr in the form of its salt, *e.g.* calcium, phosphorus, potassium, sodium, iodine, copper, molybdenum, sulphur, chlorine, magnesium, manganese, *etc.* Minerals help to build tissues, regulate body fluids or assist in various body functions. Like vitamins, minerals are required in small quantities and are vital to the body. They should be supplied daily as they are excreted through the kidney, the bowel, and the skin. They are present in the body as organic compounds such as phosphoproteins, phospholipids, haemoglobin, thyroxine, or as inorganic compounds such as sodium chloride, calcium phosphate, as free ions.

2. Fish as Mineral Source

☆ Fish salt is good source of minerals particularly small fish which are eaten with bones are good source of calcium, phosphorus, iron and copper.

☆ The only rich source of iodine commonly included in the human diet is fish Dried fish is an important part of the village diet which provides substantial amount of calcium and phosphorus.

☆ Most especially oysters are rich in calcium, magnesium, copper, zinc and iodine.

3. Function of Minerals

They are structural components of bones, teeth, soft tissues, muscles, blood and nerve cells. They help in regulating the activity of nerves with regard to stimuli and contraction of muscles. They help to maintain the acid – base balance of the body fluids. They control the water balance in the body by means of osmotic pressure and by regulating the permeability of cell membranes. They help to utilize food by helping in the process of digestion. They are part of the molecules of hormones

- **Minerals as Body Regulators:** To maintain normal exchange of body fluids (osmosis), magnesium, sodium, potassium, chlorine and bicarbonate are essential. Balance of calcium with sodium and potassium is essential for muscles contraction and irritability of nerves. Calcium is needed for clotting of blood. Iron, iodine, selenium and manganese are needed for water balance and maintenance of pH in the body fluids.
- **Minerals as structural components:** Calcium and phosphorous are needed for the formation of bones and teeth. Deficiency symptoms are stunted growth, weekend or soft bones, malformed or decaying teeth, and rickets. Potassium, phosphorus, sulphur and chlorine are needed for the maintanence of hair, nails, skin and soft tissues. Phosphorus is needed for the functioning of nervous tissue. Iron, calcium, sodium, phosphorus and copper are essential for the functioning of blood. Lack of iodine results in enlargement of thyroid glands(simple goitre).
- **Minerals needed for Glandular secretions:** Chlorine is essential for the secretion of gastric juice. Iodine is needed for the formation of thyroxine in thyroid gland. Zinc is needed for the secretion of insulin from pancreas.

Table 7.1: Minerals Content of Edible Portion of Fish

Mineral	*Average Content (mg percent)*	*Mineral*	*Average Content (mg percent)*
Potassium	300	Iron	1.5
Chloride	200	Manganese	1
Phosphorus	200	Zinc	1
Sulphur	200	Fluorine	0.5
Sodium	63	Arsenic	0.4
Magnesium	25	Copper	0.1
Calcium	15	Iodine	0.1

4. Classification

Minerals may be classified into three groups:

- Major minerals or macro minerals – these are required in large amounts, at least 100 mg/ day. e.g. calcium, phosphorus, sodium, chlorine, potassium.
- Minor minerals– these are required in small quantities, less than a few mg a day, *e.g.* iron, sulphur, magnesium
- Trace elements or micro minerals – their requirement is in few micro gram level. *e.g.* iodine, fluorine, molybdenum, zinc.

4.1. Major Minerals or Macro Minerals

These are required in large amounts, atleast 100 mg/day, *e.g.* calcium, phosphorus, sodium, chlorine, potassium.

4.1.1. Calcium

The most abundant mineral found in the skeletal and bone tissues. Approximately 1200 gm of calcium is found in our body, of which 90 percent is combined as salts – mainly as calcium phosphate-which give hardness to the bones and enable them to hold the weight of a body. The remaining 10 percent performs the following functions.

Functions

It catalyses the clotting of blood. Calcium activates several enzymes such as pancreatic lipase, adenosine triphosphatase, and some proteolytic enzymes. It is necessary for the synthesis of acetylcholine, a substance that is necessary for transmitting nerve impulses. It helps to absorb vitamin B_{12} in the ileum. Its vital function is to regulate the absorption process by increasing the permeability of cell membranes.

Deficiency

A severe deficiency of calcium leads to rickets in children and osteomalacia in adults.

Rickets

It is seen more often in premature infants than full-term infants as their growth and calcification rate demand more calcium and vitamin D. The characteristics of rickets are delayed closure of the fontanelles, bulging or bossing of the forehead and the bones are soft and fragile. There may be bowing of the legs or knock-knees, enlargement of wrist, knee and joints. Muscle development is poor and walking may be delayed. There may be restlessness and nervous irritability. Rachitic rosary is seen in the breast region as enlargement of the constochondral junction. Projection of the sternum causes pigeon breast.

Osteomalacia or Adult Rickets

Softening of the bones may be seen in women due to repeated pregnancies and low intake of vitamin D and calcium. It can also be seen in patients with renal disease. This condition is referred to as osteomalacia. Softening usually takes place in the bones of the legs, spine, thorax and pelvis, which may bend and show deformities. Other symptoms of the disease are rheumatic type pain in bones of the legs and lower part of the back, general weakness, especially while walking and climbing stairs, and spontaneous multiple fractures.

Hypercalcemia

Excessive intake of calcium can result in raising the blood calcium level, thereby causing an increased deposition of calcium in the soft tissues and increased excretion of calcium in urine. This has been observed in patients with peptic ulcers who consume excessive alkalies along with large amount of milk for a long period. It may be accompanied by symptoms such as vomiting, gastro-intestinal bleeding and an increase in blood pressure. Infants who are given excessive doses of vitamin D exhibit gastro-intestinal upsets and retarded growth. It can be corrected by a removal of the vitamin from their diet.

Sources

Rich sources include milk in any form and cheese. Good sources are green leafy vegetables such as mustard greens, spinach, colocasia, fenugreek (methi) *etc.* Small fish such as madeli, which is generally eaten with bone, is also a good source. Meat and cereal grains are poor sources of calcium. Paan (betel leaf smeared with lime) generally eaten in our country, also contributes a good amount of calcium to our diet.

4.1.2. Phosphorus

Phosphorus is a vital mineral forming the matrix of the genetic substances DNA and RNA, which control heredity. About 85 percent of it is in inorganic combination with calcium in bones and teeth. The remaining amount functions as organic compounds for the following. It is a constituent of the sugar – phosphate linkage in the structures of DNA and RNA. Phosphorus as phospholipids regulates the transport of cell solutes in and out of the cell. Phospholipids that facilitate the transport of fats in circulation. Phosphorus regulates many metabolic processes that involve phosphorylation. These include absorption of glucose from the intestines and its uptake by the cell as well as resorption of glucose by the renal tubules. Adenosine diphosphate and adenosine triphosphate system is necessary for a continuous storage and control release of energy, also in the form of nicotinamide adenine-dinucleotide phosphate and in the active form of thiamine as TPP (Thiamine pyrophosphate). The active forms of many vitamins are their phosphate derivatives, *e.g.* B_1, B_2, B_6, pantothenic acid. High energy storage compounds are phosphate esters of organic compounds *e.g.* ATP, creatine phosphate *etc.*

4.1.3. Sodium

Sodium chloride or common salt is the taste enhancer in our diet. Common salt always occupies a very important position in history.

Function

Our body contains approximately 1.8g of sodium per kg of our weight and performs the following functions. Sodium is a principal electrolyte in the extracellular fluid which maintains normal osmotic pressure and water balance. It serves as a base in the extracellular fluid. It contributes to alkalinity in the gastro-intestinal secretions. Along with other ions, it maintains the normal irritability of nerve cells and helps muscle contractions. It regulates cell permeability. It maintains electrolyte differences between intracellular and extracellular fluid compartments.

Deficiency

Sodium depletion occurs in athletes and persons engaged in heavy labour. They lose significant amounts of sodium in sweat. These losses must be replaced by eating more salt. Other instances of sodium depletion are during continuous vomiting and diarrhoea. Deficiency symptoms include weakness, giddiness, nausea, lethargy, muscle cramps and in case of severe depletion, there is circulatory failure.

Sodium Retention

In cardiac and renal failure, excretion of sodium is reduced. It is retained in the cell along with excess of extracellular fluid and edema or swelling results. Other causes of sodium retention are excessive secretion of cortical hormones by adrenal tumours; similarly, ACTH used therapeutically in a variety of conditions also increase the retention of sodium.

Sources

Table sat is the main source of sodium containing about 40 percent it. It is used in cooking for preserving and improving the taste of the food. One teaspoon salt contains about 2000 mg to 2400 mg sodium. Other sources of sodium are milk, egg-white, meat, poultry, fish and some vegetables such as spinach, beets, celery, *etc.* other sources that contain low amounts of sodium are vegetables, fruits, cereals and legumes. Drinking water generally contains low amount of sodium. The commonly used sodium compounds in prepared foods are baking soda, baking powder, monosodium glutamate (ajinomoto), sodium citrate and sodium propionate.

4.1.4. Potassium

Potassium is necessary for the human body mainly all tissue cells. Like sodium, it is also found in the extracellular fluid.

Function

Our body contains approximately 2.6 g of potassium per kg of body weight, and performs the following functions. Potassium is an obligatory component of all cells, hence the greater the number of cells, the more is the increase in potassium. Potassium is required for the maintenance of osmotic pressure and fluid balance within the cell. Sodium is found to perform the same function in the extracellular fluid. Potassium is required for enzymatic reactions which take place within the cell. Some potassium is bound to phosphate in the process of formation of glucose to glycogen. When glycogen is broken down to glucose, potassium is utilized to transmit nerve impulses along with other ions.

Deficiency

Under normal dietary intake, potassium deficiency does not occur. But, it may occur in instances of severe malnutrition, chronic alcoholism, anorexia nervosa, low carbohydrate diets and in weight reduction regimes in which food intake is restricted. Characteristics of potassium deficiency are low plasma levels of potassium (hypokalemia) and the symptoms include nausea, vomiting, listlessness, apprehension, muscle weakness, hypotension, tachycardia (increased heartbeats), arrhythmia and an altered electrocardiogram. Heart muscle may be affected so severely that it may stop during diastole (relaxation phase or heartbeat).

Hyperkalemia

Hyperkalemia or excess of potassium may occur in conditions of severe dehydration, renal failure, after very rapid administration of potassium and in adrenalin insufficiency. Hyperkalemia may be fatal since heart muscles may stop during systole (contraction phase of heartbeat). Other symptoms are paresthesias of the scalp, face, and tongue, muscle weakness, poor respiration, cardiac arrhythmia and changes in the ECG.

Sources

Potassium is easily available in foods. Good sources are meat, poultry, fish, milk and curds. Rich sources are wholegrain cereals and pulses, vegetables and fruits, *e.g.* bananas, potatoes, tomatoes, carrots, celery, organs and grapes, chiku and custard apple.

4.1.5. Chlorine

Chlorine is found in the body as chloride ion. Large amounts of chloride are found in the extracellular fluid but some amount is also found in the red blood cells, and to a lesser degree in other cells.

Functions

Our body contains approximately 50 meq of chloride per kg of body weight and it performs the following functions. In the extracellular fluid, chloride is necessary for the regulation of osmotic pressure, water balance, and acid-base balance. In the gastric juice, it is the chief anion along with hydrogen ion because of which hydrochloric acid is formed. It provides an acid medium for activating digestive enzymes as well as enabling digestion in the stomach. Chloride activates amylases.

Deficiency

Severe deficiency of chloride occurs in cases of excessive vomiting, and diarrhea. Large amounts of chloride is lost resulting in alkalosis due to replacement of chloride with bicarbonate.

4.2. Minor Minerals

These are required in small quantities, less than a few mg a day, *e.g.* iron, sulphur, magnesium and trace elements.

4.2.1. Magnesium

This mineral is found in the body in much smaller amounts than calcium and phosphorus. It is found mainly as phosphate and carbonate. Magnesium is found close to the surface of the bone. It performs the following functions in the body.

Functions

It catalyses numerous metabolic reactions. It is required to activate the enzymes involved in the oxidative phosphorylation of ADP to ATP and also for the return of ATP to cyclic AMP, which in turn regulates para hormone secretion. It is required in a balanced amount in the extracellular fluid with Ca, Na, and K so that transmission of nerve impulses and the consequent muscle contraction can be regulated. Magnesium is essential for the functioning of heart-beat and maintenance of blood pressure.

Deficiency

Deficiency of magnesium does not occur under normal health conditions and food intake, but a continuous poor intake of magnesium accompanied by increased excretion leads to a rapid lowering of the plasma magnesium. Symptoms that seen are similar to those in hypocalcemictetany, *i.e.* muscle tremor, paresthesias and sometimes convulsive seizures and delirium. Conditions of magnesium deficiency often result in chronic alcoholism, cirrhosis of the liver, kwashiorkor, severe vomiting, diabetic acidosis, and diuretic therapy. In the last two cases, loss of magnesium from the body is increased.

Sources

Dairy products (excluding butter), fresh green vegetable, meat, nuts, sea food and legumes are good sources of magnesium.

4.2.2. Sulphur

This mineral is present in all body cells especially as in the form of sulphur containing amino acid, smethionine, cysteine and cysteine. Sulphur is a constituent of thiamine and biotin, two water-soluble vitamins. Skin, hair, nails and connective tissue are rich in sulphur.

Functions

The sulphur absorbed from the intestines is mainly in the organic form as the sulphur containing amino acids, methionine, cysteine and cystine. Sulphur is structural constituent of mucopolysaccharides such as chondroitin sulphate which is found in cartilage, tendons, bones, skin and the heart valves. In enzymes, it is present as –SH group.

4.2.3. Iron

Iron is needed for the haemoglobin synthesis. Several oxidase enzymes such as catalase, cytochrome oxidase and xanthine oxidase contain iron as an integral part of their molecular structures.

Deficiency

Anaemia is found commonly in infants, pre-school children, adolescent girls and pregnant women. Characteristic symptoms are low serum level of iron, high iron-binding capacity, low haemoglobin, low red cell volume, and low mean (corpuscular) haemoglobin, small and pale cells (microcytic, hypochromic). Haemosiderosis occurs when there is abnormal destruction of the red blood cells as in haemolytic anaemia. An excessive daily intake of iron produced typical symptoms of cirrhosis of liver as the iron is stored in the liver in large amounts.

Sources

Liver, lean meats, fish and poultry are good sources of iron in the form of haeme. Legumes, dried fruit and whole grain and hand-pounded cereals as well as green leafy vegetables are good sources. Jaggery and rice flakes contain inorganic iron since they are processed in iron vessels and are a fair source. However, absorption of haeme iron is much better than that of non-haeme iron.

4.2.4 Copper

Along with iron, copper involved in the synthesis is haemoglobin synthesis. Liver, brain, heart and kidney contain high concentrations of copper.

Function

Copper-protein complex known as ceruloplasmin is the major complex of copper. Some amount is bound to albumin. The cerupoloplasmin plays an important role in the transport of iron in transferrin for haemoglobin. Other functions include taste, sensitivity, formation of melanin pigment, maturation of collagen and elastin formation, integrity of the myelin sheath, synthesis of phospholipids, and bone development. It is part of several enzymes like cytochrome oxidase, dehydrogenase, tyrosinase, *etc.*

Deficiency

Deficiency of copper is uncommon in human beings. It may be observed in kwashiorkor, nephrotic syndrome, sprue and sometimes in anaemia. An excessive dosage of copper is toxic. It leads to hepatitis, lenticular degeneration, renal malfunction and neurologic disorders. Workers in copper mines often suffer such toxic conditions.

Sources

Organ meats, shell-fish, whole-grain cereals, legumes and nuts are good sources of copper, while milk is a poor source.

4.3. Trace Elements

They are required in micro gram levels to the body, e.g: iodine, fluorine, molybdenum, zinc etc.

4.2.1. Iodine

This is a trace element required in minute quantities for our body. But this amount has a far -reaching effect on the growth and metabolism of our body.

Function

The only known function of iodine is as a constituent of the thyroid hormones, thyroxine and triiodothronin. In these hormones, tyrosine along with iodine regulates the rate of oxidation within the cells and determines the rate of metabolism.

Deficiency

People of the hilly regions suffer from iodine deficiency syndrome, namely simple or endemic goitre. The soil in these regions lacks iodine; therefore, consumption of vegetables, cereals and pulses grown here does not contribute iodine in appreciable amounts. The entire population in such areas suffers from goitre hence the name endemic goitre. Goitre is an enlargement of the thyroid gland since the size (hypertrophy) and the number of (hyperplasia) epithelial cells in the gland increase. Women are more affected than men, especially in adolescent and pregnant women it is more pronounced.

Cretinism is an iodine deficiency syndrome of foetus, which is characterized by low basal metabolic rate, muscular flabbiness and weakness, dry skin, rough hair, enlarged tongue, thick lips, retardation of skeletal development and severe mental retardation. Administration of desiccated thyroid brings relief to a certain extent in that physical growth is improved and mental retardation is less severe but damage to the central nervous system cannot be reversed and is a permanent feature of the cretinism.

Goitrogens

Substances known to interfere with the activity of thyroxine and produce goitre in spite of a normal intake of iodine are known as goitrogens. Goitrogens are found in cabbage, turnips, rutabagas, cauliflower, brussel sprouts, peanuts and mustered. However, the goitrogens in these foods are inactivated by cooking that involves high temperature.

Sources

Marine fish, shellfish and seaweed contain large amounts of iodine. The iodine content of eggs, dairy products, meat and poultry depends upon the iodine content of the animal's diet. Vegetables and fruits that are grown in soil that is rich in iodine are good sources of iodine. Hence, the iodine content of the food obtained from animals or plants depends upon the iodine in the animal's diet or the soil in which the plants is grown. Fortification of common salt with potassium iodate is a recommended method of making iodine easily available. Special government subsidies are available for preparing iodized salt. Even legislative measures making it compulsory to make iodine available through common salt are under way. This will go a long way in preventing iodine deficiency in our country.

5. Digestion and Absorption of Minerals

Most of the minerals are best absorbed by the body in their soluble form. Organic form is better absorbed by the body, than inorganic form *e.g.* iron in the inorganic form in rice flakes and jiggery is not absorbed as much as that in meat.

For a normal adult, a balance exists between the intake of an element and its excretion. Homeostasis is a mechanism by which the continuous flow of nutrients into the cell and out of it, *e.g.* bone. This tissue of the body is thought to be inert, but it is not so. It is an active tissue in its uptake and release of nutrients, especially minerals. In spite of this, there is always a state of a balance so that at any given time there is an adequate supply of nutrients. Indian diets are generally deficient in calcium, iron and iodine. If these three minerals are supplied by a diet in adequate amounts, the requirement for the rest will be easily met.

Table 7.2: Recommended or Suggested Daily Intake of different Microelements for Human Proposed by a Few Authorities

Element	*Suggested Daily Intake for an Indian*	*Recommend Daily Allowance for Healthy Daily People of U.S.*	*Estimated Safe and Adequate Intake*
Iron	11.5-49.5 mg	6 mg-30 mg	-
Copper	2.2 mg	-	2.0-5.0 mg
Zinc	15.5 mg	5-19 mg	-
Manganese	5.5 mg	-	2.0-5.0 mg
Iodine	-	40-200 μg	-
Selenium	-	10-75 μg	-
Molybdenum	-	-	75-200 μg
Chromium	-	67 μg	50-200 μg
Fluorine	-	-	1.5-4.0 mg

6. Electrolytes and Trace Elements

Ash content of muscle of a fish gives a measure of its electrolyte and trace element contents. Electrolytes are various ions which are distributed in bio-fluids of a living system including fish and shellfish. Of the different cations *e.g.*Na+, K+, Ca++ and Mg++, Na+ and K+ are the predominant ones. Important anion are chloride and phosphates. There are other anions such as bicarbonate, organic acids and proteins but they are not been taken into consideration. Many of the elements which are grouped under electrolytes and minerals are also present as inorganic compounds (in bones), as organo-metallic compounds in many enzymes and also in many other biochemically active substances.

All the above present in fish serve as nutrients to human system and physiology. However, fish as a supplier of these nutrients has received less attention from the nutritionists. Many of these elements, particularly those belonging to the group of microelements are harmful to human health if taken in excess certain limits.

Chapter 8

Marine Peptides

1. Introduction

Peptide is a molecule consisting of two or more amino acids. Peptides are smaller than proteins, which are also chains of amino acids. Molecules small enough to be synthesized from the constituent amino acids are called peptides rather than proteins. The dividing line is at about 50 amino acids. Depending on the number of amino acids, peptides are called dipeptides, tripeptides, tetrapeptides, and so on. They are are biologically occurring short chains of amino acid monomers linked by peptide bonds.

- ☆ The word peptide is derived from greek word "peptos"/"pessein" which means for digested or to digest.
- ☆ Bioactive peptides were first discovered and isolated in marine species as neurotoxin, cardiotonic peptide, antiviral and antitumor peptide, cardiotoxin and antimicrobial peptide.

2. Types of Marine Peptides

Based on the functional properties marine peptides are classified into

1. Antimicrobial Peptides
2. Antiviral Peptides
3. Antitumor/Cytotoxic Peptides
4. Antihypertensive Peptides/Angiotensin-I Converting Enzyme (ACE) Inhibitory Peptides
5. Antioxidant Peptides

6. Cardiovascular Protective Peptides
7. Immunomodulatory Peptides
8. Neuropeptides
9. Neuroprotective Peptide
10. Anti-Diabetic Peptide
11. Analgesic Peptides
12. Appetite Suppressing Peptide

3. Marine Peptides and its Role in Human Health

A. Antimicrobial Peptides

Antimicrobial peptides (AMPs) play a crucial part in the innate immunity and can be regarded as host defensive peptides.

Example

Antimicrobial peptide epinecidin-1 from grouper (*Epinephelus coioides*) demonstrated antibacterial activity against *Pseudomonas aeruginosa, Staphylococcus coagulase, Streptococcus pyogenes* and *Vibrio vulnificus.*

B. Antiviral Peptides

The marine antiviral peptides are usually found in sponges which are conventionally well-known for their unique bioactive metabolites. The bioactive peptides from sponges are usually cyclic or linear peptides containing atypical amino acids. The three main mechanisms of antiviral effects of antiviral peptides are: (i) peptides that inhibit attachment of viruses and virus-cell membrane fusion; (ii) peptides that disrupt the viral envelope; and (iii) peptides that inhibit replication of influenza virus by interacting with viral polymerase.

Example

The cyclic peptides mirabamides A–D from the marine sponge *Siliquariaspongia mirabilis* suppressed HIV-1 fusion.

The anti-HIV cyclodepsipeptide, homophymine A from the marine sponge *Homophymia* sp. possesses 11 amino acid residues and an amide-linked 3-hydroxy-2,4,6-trimethyloctanoic acid moiety and four unusual amino acid residues.

C. Antitumor/Cytotoxic Peptides

Major mechanism for antitumour peptides discovered from marine organism was inducing cell death with different mechanisms, including apoptosis, affecting the tubulin-microtubule equilibrium, or inhibiting angiogenesis.

Example

Somocystinamide A (ScA), a lipopeptide, was isolated from *Lyngbya majuscule/ Schizothrix* sp. of marine cyanobacteria.

D. Antihypertensive Peptides/Angiotensin-I Converting Enzyme (ACE) Inhibitory Peptides

Hypertension, or high blood pressure, is a chronic medical condition in which the blood pressure in the arteries is elevated. The renin-angiotensin system is one of the endocrine systems for regulating blood pressure. When the blood flow to the kidneys is reduced, renin is secreted and converts angiotensinogen to angiotensin I which is subsequently converted by ACE into the potent vasoconstrictor angiotensin II, resulting in elevated blood pressure. ACE inhibitors block angiotensin conversion and cause relaxation of blood vessels ensuing in a lower blood pressure. Thus ACE inhibitors can be used as a type of drugs for the treatment of hypertension. The specific amino acid composition of a peptide is a critical factor for its ACE-inhibitory activity. Glu, Asp, Pro, Gly, and Ala are observed in many ACE-inhibitory peptides derived from food proteins.

Example

The major peptides in four fractions (A–D) produced by alcalase hydrolysis of a protein concentrate recovered from a cuttlefish industrial manufacturing effluent with potent ACE inhibitory activity. ACE inhibitory peptides have been isolated from enzymatic hydrolysates of various sources from fish waste such as Alaska Pollock skin, sea bream scales and yellowfin sole frame protein.

E. Antioxidant Peptides

Reactive oxygen species (ROS), including singlet oxygen, hydrogen peroxide, superoxide anion, and hydroxyl radicals and other free radicals attack macro molecules such as DNA, proteins, and lipids, leading to many health diseases including cardiovascular diseases, aging, diabetes mellitus, neurodegenerative diseases, and cancer. Antioxidants are well-known for their beneficial effects on health as they can protect the body against reactive oxygen species (ROS) and free radicals which exert oxidative damage to membrane lipids, protein and DNA.

The mechanism of action of marine peptides is due to the results of specific scavenging radicals formed during peroxidation, scavenging of oxygen containing compounds or metal chelating ability and aldehyde adduction.

Peptides isolated from marine fish proteins have greater antioxidant properties than alpha tocopherol in different oxidative system. Gelatin peptides are rich in hydrophobic amino acids, hence marine gelatin peptides possess higher antioxidant activity because of the high percentage of glycine and proline contained in them.

Example

A trio of antioxidant peptides PC-1 and PC-3 were obtained from croceine croaker (*Pseudosciaena crocea*) muscle using pepsin and alcalase hydrolysis. The 582-Da peptide derived from the peptic digest of Japanese flounder (*Palatichtys olivaceus*) skin gelatin effectively scavenged reactive oxygen species and enhanced the expression of catalase, glutathione, and superoxide dismutase thus protecting membrane lipids, DNA and proteins from injury.

F. Cardiovascular Protective Peptides

The cardiovascular protective peptides include bioactive peptides which exhibit activities important to cardiovascular health, including effects on blood pressure, oxidative stress, coagulation, atherosclerosis and lipid metabolism.

Example

Two peptides P1 and P2 isolated from an enzymatic hydrolysate of *Spirulina maxima* were tested for protective action against endothelial cell activation and early atherosclerosis brought about by histamine, which mediates inflammation in endothelial cells.

G. Immunomodulatory Peptides

Marine immunomodulatory peptides that modifies the immune response or the functioning of the immune system (as by the stimulation of antibody formation or the inhibition of white blood cell activity). Some fish hydrolysates showed strong immunomodulatory effects in animals. These effects may be due to enhanced macrophage activity and lymphocyte proliferation, natural killer cell activity and cytokine regulation.

Example

Two anti-allergic peptides from enzymatic hydrolysate of *Spirulina maxima* inhibition reduced histamine release and intracellular Ca^{2+} elevation and thereby suppressed mast-cell degranulation.

H. Neuropeptides

Neuropeptides are intercellular signalling molecules that are secreted by neurons to act as neurotransmitters, modulators of synaptic transmission or hormones. Neuropeptides are key players in neural mechanisms controlling physiological and behavioural processes; for example, neuropeptides control feeding behaviour and reproductive behaviour in vertebrates and invertebrates.

Example

Protein hydroylsates or peptides derived from marine species may affect our nervous system after oral intake. They exhibit a positive effect on motivation, emotion, behavior, stress, appetite and pain management. The gonadotropin releasing hormone, which increases the secretion of pituitary gonadotropin has been detected in hagfish and Japanese anchovy brains.

I. Neuroprotective Peptide

Marine proteins and peptides suppress the development of neurodegenerative diseases like Parkinson's disease, Alzheimer's disease, and multiple sclerosis. Their neuroprotective action is brought about by the direct interaction of absorbed protein and peptide with a variety of cellular and molecular targets with enzyme/ ion channels.

Example

The neuroprotective peptide from the seahorse *Hippocampus trimaculatus* exerted a protective action on cells and prevented them from the deleterious action of which is associated with the pathogenesis of Alzheimer's disease.

J. Anti-Diabetic Peptide

The development of new therapeutic agents to improve glucose metabolism, prevent and inhibit type-2 diabetes mellitus-related complications is greatly significant.

Example

A few fish protein hydrolysates have shown in vivo glucose uptake-stimulating activity and could be used in hyperglycaemia management in addition to regular therapy.

K. Analgesic Peptides

Conopeptides from marine snail venoms have attracted much interest as leads in drug design. Cone snails have evolved a myriad of small, stable venom peptides (conopeptides) for defense as well as prey capture. They target membrane proteins of therapeutic importance, including ion channels, transporters and G protein-coupled receptors. conopeptides serve as probes for ascertaining the role of a variety of important membrane proteins in normal and disease physiology.

Example

Conotoxins from the worm-hunting cone snails *Conus nux, Conus brunneus*, and *Conus princeps*. The peptide neurotoxin Av3 from sea anemone *Anemonia viridis* is specific for arthropod voltage-gated sodium channels.

L. Appetite Suppressing Peptide

Obesity has become a serious public health problem in the developed countries, hence much effort has been made in searching for antiobesity therapeutics. Cholecystokinin and gastrin are peptide hormones linked with the satiety signal and were found to be a potential target.

Example

Some marine peptides were found to be gastrin/cholecystokinin-like or stimulate cholecystokinin release, and thus control appetite. Low-molecular-weight peptides from shrimp head protein hydrolysates have been found to be effective and promising functional food against obesity via regulation of cholecystokinin release.

4. Source of Marine Peptides

A. Marine Peptides Derived from Fish

Fish peptides accelerate calcium absorption. Under many conditions, dietary calcium become unavailable for absorption due to the formation of in soluble compounds inside the dietary tract. Fish protein hydrolsates contain hormonelike peptides and growth factors that accelerate calcium absorption. Therefore these peptides can be used in the treatment of osteoporosis and paget's diseases.

Fish	Peptide activity
Big eye tuna (Muscle)	ACE inhibitor and antioxidant
Big eye tuna (frame)	Antihypertensive
Alaska pollock	ACE inhibitor and antioxidant
Sea bream	ACE inhibitor
Yellow fin sole	ACE inhibitor and antioxidant

B. Marine deptides Derived from Lobster, Shrimp, and Crabs

Peptides	*Source*	*Activity*
Antimicrobial peptides	Blue crab hemolymph (*Callinecles sapidus*)	Highly inhibitory to gram negative bacteria
Antimicrobial peptides	Mud crab (*Scylla serrata*)	-
Antimicrobial peptides	American lobster (*Homarus americanus*)	Bacteriostatic activity against some gram-negative bacteria, and demonstrated protozoastatic and protozoacidal activity
Antimicrobial peptides (arasin-1)	Spider crab (*Hyaes araneus*)	Inhibited the growth of *Corynibacterium glutamicum*

C. Marine Peptides Derived from Squid, Clam and Sea Urchin

Peptides	*Source*	*Activity*
Antioxidant peptides	Tryptic hydrolysate of jumbo squid (*Dosidicas gigas*)	Strong inhibition of lipid peroxidation that was much higher than that of the natural antioxidant alpha tocopherol.
ACE-inhibitory peptides	Peptic digest of short necked and Clam hydrolysate	-
ACE-inhibitory peptides	Hard clam (*Meretrix lusoria*) residual meat hydrolysate	-
Antibacterial peptides Strongylocin 1 and 2	Coelomocyte extracts of green sea urchin (*Strongylocentrotus droebachiensis*)	Potent activity against gram-negative and positive bacteria.

D. Marine Peptides Derived from Mollusk

Cytotoxic cyclic and linear peptides (Dolastatin)	Marine mollusk (*Dolabella auricularia*)	Prominent cell growth suppressing activity
Compact and stable linear peptides (Conotoxins)	Mollusks	Exhibit specific actions on the ion channels and membrane receptors of excitable cells.

Anticancer peptides (Pentapeptide dolostatin 10)	Opisthobranchia mollusck	Most active natural anticancer substance
Cytotoxic cyclic hexapeptide (Keenamide A)	Marine mollusk (*Pleurobranchus forskalii*)	Possess powerful cytotoxic activity against multiple tumour types.

5. Conclusion

Marine peptides were found to be the safe and efficient agents in prevention or treatment of chronic diseases, such as heart disease, stroke, cancer, chronic respiratory diseases and diabetes. They are valuable source of bioactive compounds that could be used for the food and pharmaceutical industries.

Chapter 9

Marine Collagen

1. Introduction

Collagens are a large family of triple helical proteins that are wide spread throughout the body and are important for a broad range of functions, including tissue scaffolding, cell adhesion, cell migration, angiogenesis, tissue morphogenesis and tissue repair. Collagen is best known as the principal tensile element of vertebrate tissues such as tendon, cartilage, bone and skin, where it occurs in the extracellular matrix as elongated fibrils. Collagen is also well known for its location in basement membranes – for example: in the kidney glomerulus, where it involves molecular filtration.

Fibroblast is the most common cell that creates collagen.

The name collagen comes from the greek word (kolla) meaning "glue" and gen denoting "producing"

2. Classification

Collagen can be divided into several groups according to their structure, so far 28 collagen types have been identified.

1. Fibrillar (Type I, II, III, V, XI)
2. Non fibrillar Fibril associated collagens with interrupted triple helices (Type IX, XII, XIV, XVI AND XIX)
3. Short chain (Type VIII, X)
4. Basement membrane (Type IV)
5. Multiplexin

Multiple triple helix domains with interrupted triple helices

6. MACIT

   Membrane associated collagens with interrupted triple helices (Type XIII, XVII)

7. Other (Type VI and VII)

Collagen occurs in many places throughout the body. Over 90% of the collagen in the human body, however, is type I.

Table 9.1: The Most Common Five Types

Type	*Source*
Type I	Skin, tendon, vascular ligature organs, bone (main component of organic part of bone)
Type II	Main collagenous component of cartilage
Type III	Reticulate (main component of reticulate fibers) commonly found alongside type I
Type IV	Forms basal lamina, the epithelium –secreted layer of basement membrane
Type V	Cell surfaces, hair and placenta

3. Source

The most common raw materials used for collagen and gelatin extraction are skins or hides, bones, tendons and cartilages. Raw materials from fish and poultry have received considerable attention in recent years, but limited production makes them less competitive in price than mammalian gelatins.

Source of Gelatin

Cold water species: Such as Cod, Atlantic salmon, haddock, Alaskan pollack or hake;

Tropical or sub-tropical species: Black or red tilapia, Nile perch, channel catfish, yellowfin tuna, croaker, shortfin scad, skate or grass carp; flat species, such as sole

Cephalopods: Giant squid

Source of Collagen

Trout, hake, plaice, squid, deep-sea redfish, threadfin bream, walleye pollack, brownstripe red snapper or unicorn leather jacket (*Aluterus monoceros*), serve as good source of collagen

From scale: sea bream, red tilapia, sardine, grass carp, deep-sea redfish, Asian silver carp and lizard fish.

Unlike skins, scales are rich in compounds such as calcium phosphate and calcium carbonate; therefore, pre-treatment removal of calcium from fish scales is critical in order to obtain the final yield, purity, and gel strength of the gelatin/ collagen.

Giant red sea cucumber (*Parastichopus californicus*) has been considered as a potential source of collagen for nutraceutical and pharmaceutical applications.

4. Extraction

	Pretreatment	*Extraction*
Japanese Seerfish (*Scomberomorous niphonius*) Skin and bones	*Skin* Fat removal with 0.1M NaOH 1:10 (w/v) for two days at 4°C, Washing with cold water until neutral pH. Extraction with 10 percent butyl alcohol for 2 days.	*Skin* Extraction with 0.5M acetic acid, 1:15 (w/v) for 24 hours, followed by extraction with porcine pepsin (750U/mg dry weight) in 0.5M acetic acid, 1:15 (w/v) at 4°C for 2 days.
	Bones: Extraction with 0.1M NaOH 1:20 (w/v) for 48 h. Washing with cold water until neutral pH. Descaling with 0.5 M EDTA 2 Na for 5 days. Extraction with 10 percent butyl alcohol for 2 days.	*Bones:* Extraction with 0.5M acetic acid, 1:15 (w/v) for 3 days. Subsequent extraction with porcine pepsin (20U/g of residue) in 0.5M acetic acid for 4 days.

5. Functional Properties

Besides their basic hydration properties, such as swelling and solubility, the most important properties of collagen and gelatin are:

- Properties associated with their gelling behaviour, *i.e.* gel formation, texturizing, thickening and water binding capacity,
- Properties related to their surface behaviour, which include emulsion and foam formation and stabilisation, adhesion and cohesion, protective colloid function, and film-forming capacity and microencapsulation

Gel formation, viscosity and texture are closely related properties determined mainly by the structure, molecular size and temperature of the system. The surface properties of collagen and gelatin are based on the presence of charged groups in the protein side chains, and on certain parts of the collagen sequence containing either hydrophilic or hydrophobic amino acids.

6. Nutraceutical Properties

6.1. Antioxidant Activity

Antioxidants can protect foods and human body against deterioration by reactive oxygen species, retarding the progress of many chronic diseases. These antioxidant peptides derived from collagen may exert higher antioxidant effect than others derived from other protein sources. Hydrolysates exhibiting antioxidant activity have been obtained from collagen or gelatine of different marine sources, such as Alaska Pollack skin, hoki fish skin, cobia skin, brown-stripe red snapper skin, tuna backbones, sole skin, jellyfish umbrella, and squid skin.

Table 9.2: Antioxidant Peptides Derived from Marine Collagenous Sources

Source	*Activity*
Alaska Pollack skin gelatine (*Theragra chalcogramma*)	Inhibition of lipid peroxidation
	Increase of cell viability
Squid skin gelatine (*Dosidicus gigas*)	Radical scavenging
	Increase of cell viability
Tuna backbone	Radical scavenging
	Inhibition of lipid peroxidation
Squid tunic gelatine (*Dosidicus gigas*)	Radical scavenging
	Ferric reducing power
Nile Tilapia gelatine (*Oreochromis niloti-cus*)	Protective effect against free radical-induced cellular and DNA damage
Pacific cod skin gelatine (*Gadus macrocephalus*)	Radical scavenging protective effect against oxidation-induced DNA damage
Alaska Pollack skin gelatine (*Theragra chalcogramma*)	Inhibition of lipid peroxidation
	Increase of cell viability

6.2. Antihypertensive Activity

Some collagen and gelatine hydrolysates have shown potential to be used as mild or moderate ACE inhibitors.

Table 9.3: ACE-Inhibitory Peptides Derived from Marine Collagenous Sources

Source
Alaska Pollack skin gelatine (*Theragra chalcogramma*)
Squid tunic gelatine (*Dosidicus gigas*)
Nile Tilapia gelatine (*Oreochromis niloticus*)
Pacific cod skin gelatine (*Gadus macrocephalus*)
Pacific cod skin gelatine (*Gadus macrocephalus*)

6.3. Antimicrobial Activity

Antimicrobial peptides are mostly small cationic peptides, and are observed throughout nature.

Linear antimicrobial peptides identified have a high proportion of selected residues, particularly Pro, Arg, or Gly, which are abundant in collagen. Fractions from tuna hydrolysate seemed to be more active than squid for *Lactobacillus acidophilus*, *Pseudomona aeruginosa* and *Salmonella choleraesuis*; whereas squid fractions were more effective at inhibiting the growth of *Aeromona hydrophila*. The best antimicrobial ability was mainly found in the lowermost molecular weight fractions, especially in squid samples. Peptides from fish gelatine have a repeated motif of Gly-Pro-Ala triplets in their structure, and this hydrophobic character would let peptides enter the membrane, as the positive charge would initiate the peptide interaction with the negatively charged bacteria surface and consequently the pore formation.

6.4. Role in Bone and Joint Disease

Osteoarthritis and osteoporosis are two of the most common musculoskeletal disorders.

Fish collagen hydrolysate has a marked effect on chondrogenic differentiation of equine adipose tissue-derived stromal cells. Effectiveness of collagen hydrolysates on biosynthesis of macromolecules would be based on their unique amino acid composition. Oral administration of collagen hydrolysate from shark skin has increased production of newly synthesized type I collagen and proteoglycan in the bone matrix

6.5. Opioid-like Activity

Glyproline peptides have exhibited interesting opioid-like effects in animals. Glyproline family includes simple Pro-containing peptides widely found in gelatine and collagen hydrolysates. Some of these peptides could cross the blood brain barrier and directly affect central nervous structures involved in organism's response to stress factors. Glyproline peptides may also potentiate memory consolidation processes in the central nervous system. In particular, GP may suppress the analgesic action of morphine, and may diminish visceral pain sensitivity.

6.6. Calciotropic Activity (CGRP-like molecules)

CGRP (Calcitonin Gene Related Peptide) is a potent vasodilator neuropeptide in humans. Additional biological functions mediated by CGRP include: regulation of pituitary hormone secretion, release of pancreatic enzymes, control of gastric acid secretion, thermoregulation, decrease in food intake, bone remodeling, and prevention of complication during pregnancy. Different CGRP-like peptides have been found in hydrolysates derived from marine collagenous sources. Gastrin CCK-like molecules and CGRP-like molecules are derived from marine collagen of Cod backbones, North Atlantic lean fish skin, and Sardine heads.

6.7. Secretagogue Activity (Gastrin/CCK-like molecules)

Gastrin and cholecystokinin (CCK) are small intestinal hormones belonging to the secretagogue family. Gastrin is a gastric hormone that stimulates postprandial gastric acid secretion and epithelial cell proliferation. Gastrin/CCKlike molecules have been found in protein hydrolysates derived from sardine heads, cod heads, cod backbones, and skins from North Atlantic lean fish.

7. Effect of Gastrointestinal Digestion on Collagen Derived Peptides

Protein orally administered is enzymatically digested to their amino acid components in the gastrointestinal tract. Collagen-derived peptides too survive the gastrointestinal digestion. These collagen-derived peptides could also pass across the intestinal barrier and reach a maximal plasma concentration in 6 hours. Proline-containing oligopeptides are also particularly resistant to proteolysis in intestine and enterocytes, and this may result in their partial penetration into the bloodstream.

- ☆ Marine collagen may be an excellent source of molecules exhibiting interesting nutraceutical properties. The abundance of collagen in fish disposals, the resistance of some collagen-derived peptides to gastrointestinal digestion, and also their capacity to reach intact the bloodstream suggest that marine collagen could be an interesting source of bioactive peptides with promising applications in functional foods.

Chapter 10

Squalene

1. Introduction

Squalene ($C_{30}H_{50}$) is a unstructured isoprenoid, hydrocarbon. It is widely distributed in nature. It derived its name as it was isolated from the shark liver oil (*Squalus* spp.) The tropical deep sea shark (*Centrophorus arginatus*) contains high quantity, as high as 30 percent, squalane in its liver oil. Deep sea shark liver is perhaps the richest source of squalene and it is widely distributed in many vegetable oils like olive oil, palm oil, wheat germ oil, amaranth oil and ricebran oil.

2. History

The ancient Shoguns of Japan were the first to recognize the beneficial effects of extracts from the liver of deep sea sharks. They believed in its ability to provide strength, vigor, energy, virility and over –all good health. These extracts were considered as precious special gift.

Fisher men of the suruga Bay on the Japanese Peninsula of Izu, famous for shark fishing used to drink an oil they discovered from deep-sea sharks that was locally named "Samedawa," which means "cure all" which helped them survive the rigors of the sea. The extract gained favour in other parts of the world. It was called a miracle working shark liver oil in the Islands of Micronesia, aciete de bacalao or oil of the great fish by Spanish mariners in the West, and was recorded in China's ancient pharmaceutical book Honzokomuko.

In 1916, Dr. Mitsumaru Tsujimoto, a Japanese industrial engineer, described ten years after his initial discovery that Samedawa contained extremely large quantities of an unsaturated hydrocarbon and named the hydrocarbon squalene, due to its presence in the liver oil of sharks, from the Latin root "squalus" (shark). Dr Tsujimoto was presented the Imperial Award of Japan Academy in honour of his excellent research on squalene.

Ten years later, its structure was established as a dihydrotriterpene, and the authors posited that it might be an intermediary in the biosynthesis of steroids.

In 1934, Robinson proposed a direct cyclization of squalene to the steroid molecule.

In 1936, Nobel laureate researcher Paul Karrer described the biochemical structure of squalene for the first time. He was already recognized for his discoveries on the chemical structures of Vitamins A and E. Many years later only the structural formula of squalene was certified and fixed by Dr Calour, a Swiss chemist and Nobel Peace Prize Awardee.

3. Structure

The biochemical structure of squalene is $C_{30}H_{50}$ (C30:6n-&!2), a 30-carbon compound (polyprenyl, holding 6 prenyl groups, better known as isoprenoid or isoprene) shown in Figure 10.1.

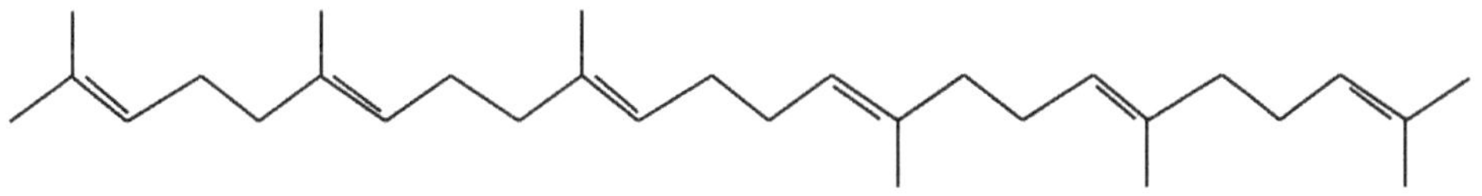

Figure 10.1: Chemical Structure of Squalene.

4. Source

Shark liver oil is considered the richest source of squalene, with squalene accounting for at least 40 percent of its weight. It is also widely distributed in nature, in lesser proportions in amaranth oil (6-9 percent), in wheat germ oil, and in olive oil (usually from 0.4 percent up to 1 percent in extra virgin olive oil). Sixty to eighty percent of dietary squalene is absorbed from dietary intake.

Squalene, as this high grade oil rich in this energy giving substance can only be found in deep-sea sharks belonging to the Squalidae group. The blue dogfish in particular contains the highest content of squalene. Basking shark contains very high quantity of squalene.

Occurrence in Indian Ocean: The deep –sea shark, abundantly available in the Indian Ocean and particularly in the seas of Andaman and Nicobar Islands, was found to contain squalene as high as 70 percent in liver oil.

Dogfish sharks live 3,000 feet under the sea. At this depth, sunlight and oxygen are almost nil. The atmospheric pressure is intense and conditions are harsh. The ability of these sharks to survive under a hostile environment is due to its large size and content of their liver which contains pure squalene. It is squalene which supplies much of the volume of oxygen needed in their bodies, providing strength and stamina.

Chemical Characteristic of Squalene

☆ Molecular weight	410.7
☆ Melting point	-75°C
☆ Viscosity of 25°C	12 centipoises
☆ Specific gravity	0.8 to 0.86
☆ Boiling point at 25°C	285°C
☆ Calorific value BTU/pound	19, 400

5. Properties

Highly purified squalene or shark liver containing over 70 percent squalene is glassy – white in colour.

The oil has a faint agreeable odour. On exposure to atmosphere squalene absorbs oxygen and colour start changing to faint-yellow and finally brown due to oxidative changes. Oxidized squalene has very little biological properties.

Squalene is a low density compound often stored in the bodies of cartilaginous fish such as sharks, which lack a swim bladder and must therefore reduce their body density with fats and oils. Squalene, which is stored mainly in the shark's liver, is lighter than water with a specific gravity of 0.855. Recently it has become a trend for sharks to be hunted to process their livers for the purpose of making squalene health capsules.

6. Synthesis

Serum squalene originates partly from endogenous cholesterol synthesis and partly from dietary sources, especially in populations consuming large amounts of olive oil or shark liver.

After its biosynthesis, squalene can be transported to other areas of the body for incorporation into tissues or it can be further metabolized, resulting in the eventual formation of cholesterol and its steroid metabolites.

7. Role of Squalene in Human Health

A. Oxygen Supply Stimulator/Detoxifier

The non-saturated hydrocarbon C30 H50, in order to stabilize, attaches hydrogen ions from water and acids in the body process, frees oxygen to the body.

The human body has about 6 billion oxygen reliant cells. Oxygenation promotes good health to the most basic level of life – the cell. Squalene helps to clean, purify, and detoxify the blood facilitating blood circulation. It cleanses the gastrointestinal tract and kidneys, causing better bowel movement and urination. Squalene revitalizes weakened body cells and helps revive cell generation. Its chief attribute is the protection it affords cells from oxidation reactions.

B. Prevent Bad Cholesterol Synthesis

Squalene were found to be more efficient in reducing total and LDL cholesterol and increasing HDL cholesterol. Though squalene is absorbed and converted to cholesterol in humans, it has not resulted in increase of serum cholesterol. Squalene is a key intermediate in the biosynthesis of cholesterol and all steroid hormones.

C. Squalene Role in Cell Membranes

Biomembranes are themselves highly vulnerable to the harm caused by free radicals, especially in the hydrophobic band between the two lipidic layers. Squalene protects the cellular and the cytoplasmatic organoids' biomembranes from oxidative stress. Squalene facilitates oxygen to reach the cellular level, causing further improvement in organ function through aerobic metabolism, which prevents acidotic cell syndrome, where cells become acidic, deteriorate, and die due to hypoxia.

D. Squalene as Anticancer Agent

The mechanisms involved in the chemopreventive activity of squalene may include inhibition of Ras farnesylation, modulation of carcinogen activation, and antioxidative activities. It potentiate anticancer drug action and offers genotoxic protection.

Squalene, usually located in this cellular zone, has an outstanding antioxidant capability recognized because of its highly stable structure. This capability is of major importance concerning the protective action against cancer (mainly thinking about reducing DNA damage). Two squalene precursors (geranyl and farnesyl) influence the synthesis and secretion of cytokines, which regulate the immune response and also regulate cellular growth and proliferation

E. Effects of Squalene on the Skin

Sebaceous glands are small glands in the skin which secrete an oily matter (sebum) in the hair follicles to lubricate the skin and hair of animals. Squalene is one of the predominant components (about 13 percent) of sebum. Squalene is not very susceptible to peroxidation and appears to function in the skin as a quencher of singlet oxygen, protecting human skin surfaces from lipid peroxidation due to exposure to UV light and other sources of oxidative damage.

G. Emollient Activity of Squalene

One of nature's great emollients, squalene is quickly and efficiently absorbed deep into the skin, restoring healthy suppleness and flexibility without leaving an oily residue.

H. Antioxidant Activity of Squalene

Squalene play a role in modulating reactive oxygen species (ROS) level, regulate antioxidant enzyme in liver, bone marrow and heart. Squalene carries oxygen in the cellular level and improves organ function and cell metabolism. It also prevents acidotic cell syndrome disease. A diet with adequate intake of oils containing squalene might be sufficient for these protective benefits, but supplementing the diet occasionally with small amounts of squalene (500 mg/d) might be prudent for individuals exposed to significant ultraviolet radiation.

I. Other Activities

Squalene is now used as immune protector and as an antiaging compound. Squalene reduces various aches and pains, helps body organs such as the kidney, liver and gallbladder to function properly, helps digestive system to function properly by reducing gastroptosis conditions, helping to shrink haemorrhoids and curbing obesity. It also acts as a relaxant, and helps to prevent various kinds of diseases and speed up the healing process in most conditions of ill health. It is also used as a drug carrier.

J. Toxicity

Squalene has not been found to have any side effect as of today. Since squalene is a naturally occurring lipid component in healthy diets, it is likely that at reasonable levels of consumption it may be safe.

Chapter 11

Effect of Cooking Methods on Fish Nutrients

1. Introduction

Nutrients are the building blocks of the human body. They enter into the cells, regulate their functions and furnish the energy for their work. Nutrients, which are provided by foods, are divided into macronutrients (proteins, fat, and carbohydrates) and micronutrients (vitamins and minerals). Nutrients may be destroyed or lost when foods are processed because of their sensitivity to heat, light, oxygen, pH of the solvent or a combinations of these. Nutrient losses may occur between harvesting as well as catering and during storage.

2. Effects of Cooking Methods on the Characteristics of Fish Lipids

Generally, cooking methods are divided into three categories: dry-heat, moist heat and combination-heat. In dry heat method, foods are cooked with the hot air or fat (Pan-frying, deep-frying, grilling, broiling, roasting, baking); moist-heat cooking methods cook the food with a liquid, usually water, stock, or steam (poaching, boiling, steaming); and combination cooking methods use a combination of dry- and moist-heat methods (braising, stewing).

3. Effect of Frying on the Characteristics of Fish Lipid

Frying can change the amount of fat content, lipid fractions, and fatty acid profiles of fish lipids. These changes mostly depend on the frying methods (shallow- or deep-frying), amount of fat content in the raw fish, and frying oil composition.

3.1. Influence of Frying Methods

Both shallow-and-deep-fat frying will change the fat content of the fish. Possible mechanisms for the changes occur in the culinary process are absorption of culinary fat into the fish, moisture loss of food, leaching of fat soluble molecules out of the food, and oxidation reactions with free radicals generated in the hot culinary fat. Deep-frying of fish induces the largest change in fish lipids due to the absorption of high amounts of frying oil, thus resulting in a drastic alternation of fish fatty acid composition. However, the effects of shallow frying are lesser due to less oil absorption in fish fillets.

Table 11.1: Contents (g/kg) of Total Lipids (TL), Polar Lipids *(PL), and Neutral Lipids (NL) in the Flesh of Raw and Fried Fish

Fish Species	*Processing*	*TL*	*P L*	*NL*
Golden Trout	Raw	57.84±6.32	31.04±7.56	17.00±0.96
	Fried	85.30±10.42	25.2±6.28	54.22±6.32
Pollock	Raw	10.67±1.78	5.83±0.34	2.83±0.76
	Fried	85.21±8.56	2.94±0.18	73.32±5.23
Seabass	Raw	30.67±2.45	16.80±2.23	11.19±1.48
	Fried	82.28±12.34	9.32±0.89	76.82±6.95
Pleuronecete Plastessus	Raw	23.56±1.78	13.27±0.96	6.55±0.41
	Fried	59.42±2.32	10.52±1.54	49.22±3.79
Haddock	Raw	20.69±4.38	10.01±0.98	7.93±0.88
	Fried	84.92±12.12	8.39±0.45	70.92±8.12
Rainbow trout	Raw	56.78±8.90	27.85±3.98	22.01±1.32
	Fried	92.95±7.05	17.93±1.02	78.03±6.92

Source: Tzortzis *et al.* (2005).

3.2. Influence of Fat Content of Raw fish

The fat content of raw fish can influence fat changes and interactions between the culinary fat and that of the fish when frying. High fat content fish resulted in lower changes in fish lipids in frying. In contrast, fish that have a low fat absorb more fat. But there is a reduction in the value of EPA and DHA in sardines and mackerel.

3.3. Influence of Frying Oil Composition

The composition of the frying oil can influence the amounts of fat content and fatty acid changes of fried fish. Changes in sardine fatty acids during deepfrying in sunflower oil, olive oil, and lard were studied. They also reported that the fat composition of the fried sardines tends to be similar to that of the frying culinary fat. Using margarine increased the percentage of SFA(Saturated FA), whereas using olive oil increased the percentage of MUFA.

3.4. Effects of Frying on the n-3 PUFA Fatty Acids

EPA and DHA, which are of much interest because of the benefits to human health, could be modified by the high temperature used in the frying process. DHA and EPA levels may reduce during frying in vegetable oil due to the oil absorption during frying.

3.5. Effects of Frying on the PUFA/SFA Ratio of Fish

The PUFA / SFA ratio is one important index to facilitate identification of healthy foodstuff. The PUFA /SFA ratio in raw sardines was raised more than 3 times when fried with sunflower oil and decreased by 47% when fried with lard. According to more recent opinions of nutritionists and physicians, the ideal diet should consist of one third SFA, one third MUFA, and one third PUFA. If this goal is achieved, the ratio of PUFA /SFA in dietary lipid should be 1, and PUFA content should be 10% of energy intake. Vegetable oils rich in n-6 PUFA should be avoided in pan and deep fat frying.

3.6. Effects of Frying on the n-6/n-3 Ratio of Fish

The amount of n-6/n-3 fried fillets was related to the quality of the frying oils. The deep fat frying of fish leads to an increase in the total amount of fat but also to an increase in the n-6/n-3 PUFA level of raw fish. The World Health Organization (WHO) has recommended that the daily ratio of n-6/n-3 in total human diet should not be higher than 5.

4. Effects of Cooking Methods Other than Frying on Fish Lipids

4.1 Oven Baking

During oven baking, silver catfish fillets lost water with a consequent increase in protein, fat, and ash content. Oven reheating minimally affected the sardine fillet fatty acid content in fried frozen sardine fillets as compared to microwave reheating. This effect must be primarily a consequence of the low fat loss product by this process. Pre-fried breaded black Pomfret fillets cooked in microwave oven and frying showed least changes in fat content and fatty acid composition.

4.2. Microwave Cooking

The effects of cooking with out oil on the FA concentrations of fish flesh are dependent upon the type of fish. They reported that the total FA contents were slightly increased without great differences between individual fat in lean pike in conventional and microwave baking methods. However, the total amount of fat was slightly decreased (5%) in oven baked and increased (32%) in microwave cooked rainbow trout samples. Microwave cooking significantly changed the total fat content in the muscle.

4.3 Grilling

Grilling of blue eye resulted in an increase in total lipids and n-3 PUFA (on a wet basis), presumably due to the decrease in tissue water content. Boiling, baking, or grilling marginally affected the silver catfish fillets' fatty acid content. It must be a consequence of the water loss produced these processes. Grilling produced higher water losses than oven baking, but lower than frying. These modifications are related to the rate of change in food temperature and the process temperature (higher in grilling than in oven baking). An increase in the MUFA and PUFA concentrations were found in boiled rainbow trout. A significant increase in the total fat content was observed by during grilling of the *Scomberomorous guttatus* muscle.

5. Effect of Canning

Influence of Canning on Food Quality and Nutritive Value

Thermal processing is likely to bring out some undesirable changes in the food. Thermal processed foods are known to undergo degradation in colour, flavour, texture and nutritive value. Heating can bring out irreversible changes in several of the components of the foods such as proteins, fat and carbohydrates.

5.1. Changes in Colour

Some change may occur in the degradation of the natural colours of thermally processed foods. Heating in general causes degradation of the natural colour, but it becomes more critical when heated in a complex substrate such as canned food. Degradation of pigments will be accelerated by the presence of metallic ions in the food. The natural carotenoid pigments in shrimp, astacene, astaxanthine, undergo some degree of degradation during canning.

5.2. Flavour and Texture

Flavour and texture of the food will get degraded as a result of heating and the extent of degradation will depend upon the sensitivity of the food to heat and the duration of exposure to heat. Heating has profound influence on the flavour-bearing components of the food.

5.3. Protein

Proteins are denatured by heat. On denaturation, the configuration of the native protein molecule gets changed simultaneously losing the specific immunological properties. Enzymes become completely inactivated. Free sulphydril groups increase and the heated protein tastes different. The denatured proteins coagulate and finally precipitate. This adversely affects the digestibility of proteins and thus the nutritive becomes impaired even though there is no change in the amino acid composition.

5.4. Fats and Oils

Fats and oils are susceptible to hydrolytsis as well as oxidative rancidity during canning. The lypolytic enzymes produced by micro-organisms are destroyed by heating and hence enzymatic hydrolysis is not a problem in canned foods.

Oxidative rancidity produces hydroperoxides and is accelerated by heat, metal ions and moisture. The rate of oxidation of fat is doubled for every 10°C increase in temperature. Flavour reversion in de-odourised unsaturated fats is accelerated by heat. Fats are stable in canned foods when there is good vacuum in the can.

5.5. Carbohydrates

Carbohydrates are degraded by heating when they enter into browning type reactions or by getting caramelized at high temperatures.

5.6. Vitamin

Fat- soulble vitamins A and D are relatively stable to heat, but appreciable loss to vitamin A will occur if heated in the presence of oxygen. In the case of vitamin, heating in the presence of oxygen results in its rapid destruction. Vitamin E is stable to heat in the absence of oxygen, but heating in the presence of oxygen leads to its almost complete destruction.

Water-soluble vitamins such as thiamine and ascorbic acid are very heat sensitive whereas riboflavin is stable to heat, but sensitive to light. Thiamine may be retained to some extent in acid foods, because of their lowheat processing requirements. Ascorbic acid gets destroyed by heating at low temperatures for longer periods, this being the combined effect of heat, metal ions, oxygen etc. High temperature short time processes are less destructive processes of equivalent lethality.

- ☆ The nutrient contents of fish vary according to the species, season, and environment conditions. The fish nurient characteristics are changed by different cooking processes. High temperature exposure produces higher changes in fish nutrients than other cooking methods. Changes during frying depend on the fat content of fish, frying oil composition and types of frying technology. Frying results in a higher loss in the DHA and EPA than other cooking methods.

References

Appel, L. J., Miller, E. R., Seidler, A. J., and Whelton, P. K. (1993). Does supplementation of diet with fish oil reduce blood pressure? A meta analysis of controlled clinical trials. *Archives of Internal Medicine,* 153, 1429–1438.

Anderson, J. J. B., and Garner, S. C. (1996). Calcium and phosphorous nutrition in health and disease: Introduction. In J. J. B. Anderson and S. C. Garner (Eds.), Calcium and phosphorous in health and disease (pp. 1–5). New York: CRC Press.

Aminullah-Bhuiyan AKM, Rathayake WMN, Ackman RG (1993) Nutritional composition of raw and smoked Atlantic mackerel (Scomber scombrus): oil and water-soluble vitamins. *J Food Compos Anal* 6(2):172–184

Benkajul, S., and Morrissey, M. T. (1997). Protein hydrolysates from Pacific whiting solid wastes. *Journal of Agricultural and Food Chemistry,* 45, 3423–3430.

Gokoglu N., Yerlikaya P., Cengiz E. (2004). Effects of cooking methods on the proximate composition and mineral contents of rainbow trout (*Oncorhynchus mykiss*), Food Chemistry, 19–22.

Gopakumar, K. (2002). Text book of Fish Processing Technology, Indian Council of Agricultural Research, New Delhi.

Kadler K. E., Baldock C., B*ella J. and Boot-Handford R. (2007). Collagens at a glance.* Journal of Cell Science, 120, 1955-1958.

Kim S. and Mendis E. (2006). Bioactive compounds from marine processing by-products – *A review, Food Research International,* 39: 383–393.

Mohanty *et al.* (2014). Amino Acid Compositions of 27 Food Fishes and Their Importance in Clinical Nutrition. *Journal of Amino Acids.*

Olu M. and Adediran A. Z. (2015). Protein Evaluation of Foods, *International Journal of Nutrition and Food Sciences*; 4(6): 700-706.

www.fao.org

Venugopal V (2009) Marine products for healthcare: functional and bioactive nutraceutical compounds from the ocean. In: Mazza G (ed) Seafood proteins: functional properties and protein supplements. CRC Press, Boca Raton, pp 51–102.

Venugopal V, Shahidi F (1996) Structure and composition of fish muscle. Food Rev Int 12(2):175–197 www.fao.org/wairdocs/tan/x5916e/x5916e01.htm.

Zen P. (2005). Advances in Fish Processing Technology, Allied Publishers Private Limited.

Zheng *et al.* (2011). Antitumour peptides from marine organisms, Marine drugs, 2011, 9, 1840-1859.

www.ingramcontent.com/pod-product-compliance
Ingram Content Group UK Ltd.
Pitfield, Milton Keynes, MK11 3LW, UK
UKHW021956270726
14060UKWH00002B/545

9 789388 173230